Critical Care

Critical Care

A NEW NURSE FACES DEATH, LIFE, AND EVERYTHING IN BETWEEN

Theresa Brown

harperstudio

An Imprint of HarperCollins*Publishers*

HarperCollins books may be purchased for educational, business, or sales promotional use. For information please write: Special Markets Department, HarperCollins Publishers, 10 East 53rd Street, New York, NY 10022.

For more information about this book or other books from HarperStudio, visit www.theharperstudio.com.

FIRST EDITION

Designed by Eric Butler

Library of Congress Cataloging-in-Publication Data

Brown, Theresa.
    Critical care : a new nurse faces death, life, and everything in between / Theresa Brown. — 1st ed.
        p.   cm.
    ISBN 978-0-06-179155-0
    1. Brown, Theresa. 2. Intensive care nursing—Biography. I. Title.
    [DNLM: 1. Critical Care—Personal Narratives. 2. Oncologic Nursing—methods—Personal Narratives. 3. Nurse's Role—Personal Narratives. 4. Terminal Care—Personal Narratives. WY 156 B881c 2010]
    RT120.I5B76   2010
    616.02'5092—dc22
    [B]
                                                            2009046576

10 11 12 13 14   OV/RRD   10 9 8 7 6 5 4 3 2 1

For
M., S., C. & A.
and
for nurses everywhere

Contents

Acknowledgments

My first and biggest thanks go to my editor and publisher, Bob Miller, who really is the greatest thing since sliced bread. Also thanks to Lynn Johnston, my agent, who has the business sense I lack and is hard-nosed and kind in all the right proportions. My colleagues at the *New York Times*, Tara Parker-Pope and David Corcoran, are both great editors. My friend Lawrence Downes helped me connect with them and had helpful talks with me about writing. And I also thank the great group at Harper Studio: Debbie Stier, Sarah Burningham, Julia Cheiffitz, and Katie Salisbury.

Friends near and far gave me unfailing support and enthusiasm. Thanks to Judith and Daniel; Jonathan and Susan; Nathan and Julia; and Mari, Kirstin, Joann, and Jessica. Thanks also to my family: my mom; my dad; my brother, Chris; and my (now former, but still in my heart) sister-in-law, Myra. All of you have encouraged me to "keep writing," and I really appreciate it.

Josh Rubin, M.D., helped answer my surgical questions, and Tim Gillison, M.D., made sure my chapter on chemo was correct. Benedito Carneiro, M.D., offered me his unwavering enthusiasm, and Barry Lembersky, M.D., was a big supporter of my writing at work.

I have worked with so many wonderful doctors that I can't name them all here, but in addition to the doctors listed above, a short list would include Sonika Gill, Rebecca Silbermann, Ajlan Atasoy, Usha Malhotra, Dhaval Mehta, and Bruce Hough. Nurse practitioners and physician's assistants have also taught me a lot, especially Cheryl Tompkins, Melissa Loucks, Cheryl Devitt, Ann Stewart, Kathy Curran, and Amy Federoff. And, of course, our attendings in oncology, the intensivists, cardiologists, and surgeons, and the rotating roster of interns and residents—all are part of the team.

Then there are all the other people who make my work in the hospital possible. First, the pharmacists, whose breadth and depth of knowledge about drugs make them invaluable. Next, the nurse's aides, who work hard at a thankless job, technicians and therapists of all stripes, transporters, dietary, housekeeping, social work, and all the other ancillary staff who keep the hospital humming.

Our patients are the most amazing people I know. They have the hardest job in the world: trying to stay alive while fighting cancer. I learn from them and find them inspiring every day. Even though none of them are mentioned by their own names in this book, without their stories I would not have had a story of my own to tell. It's a privilege to help them with their struggle.

The bulk of thanks goes to my husband, Arthur Kosowsky, and our three children. Arthur has been my number one fan, tireless editor, and emotional support system. When I got the contract for this book, he said, "I always knew you'd pull something like this out of your hat." Wow—thanks for having such faith in me. And the kids, now ages thirteen, eleven, and eleven (the twins), you three are the light of my life and the reason I became a nurse. If not for you, this book would not exist, and my entire life would be poorer. Thank you for being the wonderful people you are.

Finally, the nurses I work with are amazing, every one of you, and I thank you for all the help you gave a rookie; I wouldn't have made it without you. Mary B, my first preceptor, shared her wisdom and warmth with me. The nurses who introduced me to the stem cell side of oncology are my coworkers now. Sharon, our manager, and Sara and Lorraine, our clinicians, keep the floor running. Lettie was my preceptor, and I couldn't have asked for better. Elaine, Julie, and Dave are the experienced hands on the floor and my role models. The night shift team, who sacrifices sleep to take care of our patients in the wee hours, deserves a special note of thanks: Meghan, Wendy, Martha, Dixie, Cheryl, Rebecca, Sue, Todd, and Marianne. The rest of the floor staff, most of whom "flip" from days to nights, includes Abby, Sarah, Mary W., Mary J., Alyssa, Caitlin, Angela, Patti, Jenn, Leslie, Jan, Cara, Sandy, Becca, and Lisa. To Christi, Shannon, and Ketah, who listen, make me laugh, and keep my feet on the ground, I say, let's keep sharing meals, and keep on talking. Working with all of you makes even the hardest days better. Thanks for letting me share in your work and for showing me what good nursing is all about.

Author's Note

While all of the stories in this book are true, in an effort to protect patient confidentiality and privacy, descriptions of patients and staff, and many of their identifying characteristics, have been changed. This book is not a medical handbook and should not be used as such. All medical information is presented as correctly as possible and all errors, medical or otherwise, are mine alone.

Critical Care

ONE

Why the Professor Became a Nurse

"You left teaching English for *this*?"

I've been asked the question so many times by so many people that it no longer surprises me. After all, who in their right mind would give up being an English professor who taught writing at Tufts University to become a nurse?

Other versions of the question are no more complimentary. A favorite of mine is "You couldn't get a job, right?" And sometimes they really give me pause, like when another nurse asked me, "Why? Because you hated having summers off?" I hadn't looked at it quite that way before, and the question made me stop and wonder whether I really was crazy, since people ask when they hear my story, "Are you crazy?" I'm not, but I made a midlife career change that many people, including a lot of nurses, do not understand, and certainly would not have made themselves. The why of my decision at times eluded even me. Nursing just felt right, but I don't think even I fully understood my career change until the last night of the very last shift I would ever work as a nursing student.

That night an eleven-year-old leukemia patient who had a fever arrived on my floor at our children's hospital. I had decided to do my senior clinical at Children's because I wasn't sure if I wanted to work with kids or adults. In some ways I loved it there, but caring for kids when my own children were still young was hard, and ultimately I only applied for jobs with adult patients. Still, there I was on my last night at Children's with a new admission, a kid who'd been in and out of the hospital many times, at ten o'clock.

The patient, Sean, and his dad came up from the ED (emergency department). They talked and joked with each other, started watching movies on the TV in their room right away, and passed an enormous bag of potato chips back and forth. I got the impression they were trying to convince us, and more importantly themselves, that an impromptu hospital stay could be fun if you just had the right attitude.

Other nurses on the floor had warned me that this family was "difficult," but they seemed OK. Sean's dad had a bad back and asked a few times for more pillows since he would be sleeping on the chair in the room that folded out into a bed. I'm not sure why, but pillows are a rare commodity in hospitals. I searched both wings of the floor until I found some for him—even with a healthy back, those chair beds are not too comfortable. Sean, testing out some preteen behaviors, could be rude, so I teased him about saying "please" and "thank you" as I handed over cartons of apple juice. I described him to the resident as "cheeky," but I liked him.

They'd ordered fluids for him and antibiotics—lots of antibiotics—and Sean and his father were concerned I was going to wake them repeatedly during the night since I would need to administer one drug after another. I told them I would do my best to let them sleep undisturbed—peaceful sleep is another rare commodity in hospitals, and it's important for healing as well as peace of mind.

Still, they finished the first movie and moved onto another, until finally around two o'clock in the morning they both fell asleep. They had turned off the TV and the light. Sean's father had fallen asleep first, and then Sean, who'd been lying in the dark hospital room with his eyes wide open, keeping his thoughts to himself, dropped off to sleep, too. I went into the dark room and hung the drugs I needed to administer as quietly and quickly as I could without turning on a light. I had promised not to disturb them, and I meant to keep that promise.

Around 4:00 A.M. my preceptor, the nurse supervising me, told me Sean wanted a Tylenol. I went to see what was up. As soon as I walked into his room, he looked up at me in the darkness and said, "It feels like I can't breathe. My chest hurts." Alarms went off in my head, and I truly pictured myself as Tom, the cat in the *Tom and Jerry* cartoons, with little mallets alternately striking on opposite sides of my head, which had become one big metal bell. Oh, gee, that sounds bad, I thought to myself. What am I going to do about that? But then I did the things I most needed to do: made sure he could breathe and called the resident to tell him about Sean's change in status.

I told my preceptor, Paula, what was up, and she told me to get a set of vitals. Hearing that, I felt stupid. I had gotten so used to taking vital signs—blood pressure, heart rate, respiratory rate, and temperature—that I had forgotten they matter, that in a situation like this the patient's vital signs could give us valuable information about just how bad off he was. A low blood pressure and high heart rate would tell me he was in danger of being septic and going into shock. If his oxygen level was low, I would know that his breathing difficulty had something to do with not getting enough oxygen into his lungs.

I grabbed the equipment to take a set of vitals, but when I got back to the room, I had to wait—Sean needed to go to the bathroom. I helped him walk around the bed with his IV (intravenous) pump, and halfway there, between the bed and the bathroom, his knees buckled. He cried out, "I can't see! I can't see!" I held him up, then picked him up and somehow got him into the bathroom and onto the toilet. While I was holding and carrying him, I wondered, a little angrily, why his dad wasn't helping me. Could he really sleep through all this? I wondered, because he did look asleep, even though we must have been loud in that small room.

Once I got Sean settled on the toilet, I took his blood pressure twice. I took it with the machines we have, and I took it manually, by pumping up the cuff myself and listening for the flow of blood. Taken both ways, on both arms, his pressure was 70 over 30, much too low. He wasn't complaining anymore about loss of vision or not being able to breathe, and by this time the resident and the intern—the doctors in training who were taking care of him—were both in the room. The poor kid had to sit on the toilet while we all stood in the dark and talked about him. When I checked to make sure he was safe sitting on the toilet by himself, he yelled out, "Can't a man take a crap in peace?"

I most remember a swirl of activity. The resident, the more senior M.D. in the room, asked me to tell him exactly what happened when Sean said, "I can't see," and fell to his knees. I told the resident the story. He seemed stressed, or maybe I was just projecting my own feelings onto him. I'd seen sick people, sure, and sick kids, but never anyone who was this fragile, and the nighttime and the darkness of the room gave the whole situation a surreal feeling. We increased the rate of Sean's IV fluids because upping the amount of fluid is usually the first course of action when patients are hypotensive. Putting more fluid in the veins is an easy way to increase blood pressure and cardiac output.

However, we couldn't get Sean's pressure up, and the doctors were worried that he was going septic. The resident called in the fellow (an M.D. training in the hospital's fellowship program) from the PICU (pediatric intensive care unit), and they talked over Sean's symptoms. The doctors asked me to keep taking blood pressures, but Sean never climbed much above his early low. I watched all this with only a vague understanding of what was going on. The resident and the fellow had a couple of huddled negotiations in the dark hallway; the fellow made a few phone calls, then they told me Sean would be sent to the PICU, that he needed the more intense technical support available there.

Things calmed down while we waited for the call to transfer Sean to intensive care. Around 5:30 in the morning I went to check on him. The room was still dark, Sean's father was still asleep, and I hoped Sean had fallen back asleep as well. But he was awake, and he had some questions. His earlier cheekiness was gone, and I found myself confronted by a very scared eleven-year-old boy.

"Why did my chest hurt?" he asked.

I sat down on the edge of his bed. Answers and pieces of answers swirled around in my head, but the truth was I really didn't know, and the doctors didn't either. So I told him that. "I don't know," I said. "Maybe because you were having trouble breathing, your chest tightened up."

He nodded, then asked, "How come I couldn't see?"

I didn't have an exact answer to that question either, but I did my best, slowing my answer down to match his deliberate pace. "It could be because your blood pressure was so low," I told him. "Maybe that's why you fell, too."

He nodded again, then asked his last question, the hardest: "Why couldn't I breathe?" Every answer I thought of to this question seemed overly technical, but also just inadequate. I'm sure his breathing trouble had something to do with his low blood pressure, but I didn't have enough experience to know if hypotensive patients often felt short of breath. I couldn't make sense of it in a simple way, so I told him that, too. "I'm not sure," I said. "That's why they're sending you to the PICU, so they can find out why you felt like you couldn't breathe."

To me, all my answers sounded lame, but Sean seemed to find them comforting. He didn't want a physiology lecture or a detailed explanation of why patients who are septic drop their pressure; he wanted to ask someone his very troubling questions, and he wanted answers that made sense to him. As soon as I finished giving my explanation for the third question, he stopped talking and relaxed back into bed.

This brief conversation could be categorized as "patient education" in a nurse's note, but Sean and I exchanged something more substantial than information when we talked. He found a way to ask, "What's wrong with me?" and "Am I going to die?" And I told him, "I don't completely know, but whatever happens, I am here with you." There we were, nurse and patient, talking quietly in a dark room, confronting the vagaries of life and death. For me, this moment finally put to rest any questions I had about why I quit being a professor and became a nurse instead.

Around 6:30 that morning the staff initiated the transfer to the PICU. My preceptor had taken over for me at this point. As a student I wasn't allowed to take Sean up to the PICU on my own. Before they left, though, I went back into the room to say good-bye and to see if Sean had any more questions.

When I got to the room, I saw Sean's father, and my question about why he had not helped me during the night was answered. His face held such an intense look of distress that I wanted to look away. His eyes were hollowed out, almost sunken, and he stood there, stooped and silent, as if his only choice in life was to keep receiving blow after blow and hope he could stand it. This is love, I thought, and all the agony that love can bring. "Do you have any questions?" I asked him gently. "Do you understand what's happening?" That was all I could offer him. If only I could have wiped the slate of his face clean, taken the pain that was driving his shoulders in and down and thrown it out the window, but unfortunately I do not have that power. Sean's father would have to bear this burden himself.

In answer to my questions, he shook his head no and stood silently in the room while the doctors and I talked to Sean, and Paula got him ready to go.

I looked at my watch and realized I needed to hang feeds for one of my babies. This little guy had been born with a multitude of birth defects, and he got his food through a tube in his stomach. New bags of food have to be hung at specified times, and his was due now. I felt reluctant to leave Sean, but he was in good hands, and I had to take care of another patient. I told Paula where I was going and went to get the new feeds out of the refrigerator. I went into the room and hung the new bag, checked the baby's diaper, and threw the old bag of feeds away.

As I was leaving the room, another resident, one who had overheard me complain about Sean being cheeky, walked by and asked how he was doing. "Oh, he went to the PICU," I told her.

"What?" she said, looking genuinely surprised.

"Yeah," I said. "We just couldn't keep his pressure up." I marveled at how this phrase, which I had never before used in my life, came out of my mouth. Did I just say that? I thought to myself. Wow, I sound like a nurse.

She stopped for a minute, then said, "Good job," and kept on walking down the dark hall.

By eight o'clock Sean had gone off to the PICU, and I never saw him again. I ended up overstaying my final shift and missed out on a celebratory breakfast with my friends. I have no idea how things turned out for Sean, but I hope he and his dad are still watching movies and eating potato chips, and will be for many years to come.

When I finally got home that morning, much later than I intended, and so exhausted that sleep simply meant giving into gravity, it hit me that it was my own son's eleventh birthday. My child's biggest worry that day was, "Is Mom going to be awake enough to make me a birthday cake?" Compared to Sean's biggest worry, my son's might seem irrelevant, but I didn't see it that way, and the contrast between the concerns of those two eleven-year-old boys reveals part of what I love about nursing. Doctors diagnose, treat, and prescribe—work central to healing—but nurses really do tend to the whole person. A birthday cake in its own way is as important as getting answers to scary questions about not being able to breathe. Explaining human physiology in a dark room in the middle of the night and making birthday cakes both capture the essence of nursing: combining technical skill and knowledge with love.

After having my son, I realized I wanted a job where I was expected to care about people, not instruct and grade them. Then I had my twin daughters, and my world turned inside out and upside down with the physical challenges of the pregnancy. The midwives who helped me through the pregnancy left a lasting impression, and when I mentioned my admiration for them, a friend who's a nurse told me, "You could do that job."

It had never occurred to me before. People like me go to medical school, I thought. They don't become nurses. At that time I knew very little about what nurses really do, but my friend, who's also named Teresa, persisted. She was beginning a doctoral program in nursing, but she'd put in her time as a nurse practitioner, providing gynecological care to underserved teens in high school clinics. She talked about hospital nursing, how there were floor nurses who could "kick my ass" and what a huge advantage it was to be smart as a nurse. She also talked about patients in general, and without using the specific words, described for me the nurse's role as a "patient advocate," a phrase like "to serve and protect" or "first do no harm" that is so integral to the job that it can be considered a professional mantra.

I was hooked. Just days after that conversation I decided to go to nursing school. Six years later, when my twins were eight, I got my bachelor's in nursing degree, and a few months after graduation I passed the licensing exam to get the coveted R.N. It's a long story that involved my starting at the University of Pennsylvania, withdrawing when my husband's job in New Jersey imploded, and starting over again at the University of Pittsburgh after we moved there. During all that time, while I took science prerequisites at Rutgers, after beginning at Penn, and then enrolling at Pitt, my dedication to nursing never wavered. The more time that went by, the more I knew I had finally found a job that fit.

I tell people that "having kids changed my life," and truly if I had not had children, I would never have become a nurse. So, how I got from the ivory tower to the hospital, from English to nursing, flows directly from my becoming a mom. Pregnancy and motherhood can feel alternately like a slog and a wondrous journey. With my twins I got a double dose of that agony and ecstasy, and I found it enriched my life when I had not known it was impoverished.

Contemporary writer Frank Bidart has a great two-line poem called "Catullus: Odi et Amo":

I hate *and* love. Ignorant fish, who even
wants the fly while writhing.

I first read this poem in a college poetry class and thought it said something so true about the kind of romances I tended to find myself in: wanting the men who made me miserable. Now, married, in my early forties with three kids, I see a deeper meaning. At this moment in my life, the poem describes the kind of work I want to do and why I want to do it. I love my kids, but like Sean's father I know how painful and fraught that love can be. I care deeply for my patients, and I loathe their suffering and disease. Patients love the idea of being treated and cured, but they hate how those treatments can wrack their bodies more horribly than their disease ever did. I love the idea of helping patients, even when I don't know exactly what's wrong with them.

It's a simple enough idea: love what you do, even when you hate it. I never felt that way about being an English professor or even a teacher. I liked teaching, and at times I found it enjoyable enough, but I never felt passionately about it, for better or for worse.

So I gave up my summers off, and now I have to be at the hospital at 7:00 A.M. and work for twelve hours with no promise of a real break. But nursing stresses me out in a completely different and oddly more tolerable way than working as a professor ever did, I think because I find it so much more meaningful. Working as a floor nurse is messy and stressful, but I wouldn't exchange it for a dream classroom full of well-read, hardworking, intellectually curious college students—not in a million years, not ever. For where else can I go to sample daily the richness of life in all its profound chaos? Where else can I comfort a cheeky eleven year-old boy who has to confront his own mortality earlier than any of us ever should?

This book is the story of how I learned to do a job I love and hate, and why I keep on doing it. The "writhing" that Frank Bidart described in "Odi et Amo" is just part of it, too—no one can fight for their life without having some suffering mixed in, at least not the way we practice medical oncology right now. And that's where nurses come in. Doctors heal, or try to, but as nurses we step into the breach, figure out what needs to be done for any given patient today, on this shift, and then, with love and exasperation, do it as best we can.

Getting My Feet Wet

I was just a few weeks into orientation as a new nurse, working closely with my preceptor, and we had set for ourselves the task of moving a patient, Jim, from his bed to a chair. It sounds simple. Physical therapy had asked us to move him, and once we got a few other things done, my preceptor, Penny, gave the word.

She put a sheet over the chair, and we got into position on either side of him. He was a big guy, in his seventies, on oxygen, weak, and not really understanding much about where he was or why. We helped him stand up, pivot, shuffle his feet to cover the few inches of floor separating the bed from the chair, then lower himself down. Penny was on his right, and I was on his left, and as he sat down, we heard a loud ripping noise. It was such an odd noise that I almost dismissed it, but Penny was on alert right away.

"What was that?" she asked, her eyes wide. "What was that noise?"

I was up close to Jim, and looking behind his left side I noticed that the clean sheet we had put on the chair was now stained a pale brown, as if someone had spilled Coke on it. This, too, I almost dismissed. Where could that have come from? I wondered. The sheet had been clean and white; I knew that.

I cautiously peered around Jim's shoulder to look at his back. There I saw a huge opening, like a gash, extending down in a clean line probably about eight inches and extending into his back about three or four. I could see the different layers of tissue I had learned in anatomy class, all perfectly aligned. "His back split open," I said.

"What?" Penny asked.

"His back split open. His back split open." This was all I could say. In my experience, backs do not split open, and what I really wanted to say over and over again was "Holy shit!" Jim did not have sutures or a healing incision; his back had been intact skin just one minute before. There had been a scar, sure, from a surgery a few months before, but a healed scar. Skin is not supposed to break open from the inside, and I was having trouble processing what had just happened.

Penny stood up and moved over to the doorway. She looked at me. "Should I call it?" she asked, her face reflecting my own confusion, and possibly a more intense sense of panic than I was feeling.

"Yes," I said. "Call the code."

So she called the code using the portable phone each floor nurse carries, while I stayed in the room and held Jim on the chair. We hadn't gotten him settled all the way back into the seat, and if I hadn't been there, he would have fallen onto the floor. The code team came—the ICU nurses, the intensivists (M.D.s who specialize in intensive care), respiratory—and Penny answered their questions while I held Jim up. I couldn't see what was happening, and I couldn't hear much. Jim was heavy, and it took a lot of concentration to keep him in the chair. It was my first code, and I spent most of it standing with one knee bent and the other straight out behind me, a stance I had learned twenty years before in a women's self-defense class, now being used to keep my patient safe.

One of the intensivists, Dr. Sutherland, came over to me and asked, with just a hint of impatience, "Can he stand up?"

"No," I snapped, "he can't stand up."

Dr. Sutherland is a physician of few words, and he rations them with care. He nodded, almost to himself, and stepped back.

Then someone relieved me. I don't remember the details of who exactly came into the room and took over my job, but I do remember a sudden feeling of immense freedom. I could move, I could hear, and I could talk. Up till then, the events of the code seemed to have all been taking place behind a thick pane of glass. Being released from my task of holding Jim meant that the glass disappeared, and I could be part of the conversation about my patient.

Immediately, one of the doctors, Jim's intern, asked me how big the opening in Jim's back was. "Big," I said. "You could put your fist in it." And I wasn't exaggerating.

She looked intensely curious. "Do you want to go look?" I offered, acting briefly as tour guide to the view of Jim and his amazing split back.

She did want to look, and a couple of the other new interns came to look as well. These doctors, I was glad to see, marveled at Jim's back the same way I had. The intensivists didn't seem nearly as impressed, but while they were deciding what to do, someone noticed a large puddle of brownish fluid on the floor and asked, "Did that come out of his back?" When I said that it had, a moment of surprised quiet rippled through the room before the conversation continued.

By this point Penny was sounding almost apologetic. "Well, his respiratory status was already pretty compromised," she said, "and now it's gotten worse."

Respiratory status? I wondered. This guy had an eight-inch opening in his back, a clean incision offering a view of living tissue in all its multilayered glory. "Get this guy to the OR!" I wanted to scream, but it wasn't that simple.

The patient had a seroma, a fluid-filled sac that can develop at the site of a previous surgery. Several months before, he had a mass removed from his back, and that removal must have led to the accumulation of enough fluid to split open his back, soak the sheet he was sitting on, and splash in a big puddle on the floor. When a large enough mass is removed from anywhere on the body, it leaves a cavity or space. That space will fill with fluid as part of the healing process, but too much fluid makes it impossible for the tissue to regranulate, or regrow. To prevent that, surgeons often place drains in wounds to limit fluid collection.

I don't know if Jim had a drain placed following the initial surgery, but regardless, the large cavity in his back accumulated more and more fluid over time until the pressure became too great for his body to bear. Something had to give. The pressure from the seroma overcame what surgeons call the tensile strength of the skin incision, causing the patient's back to dramatically, if painlessly, rip apart. Moving him from the bed to the chair probably contributed to the bursting of the seroma. We would have pulled his arms forward to move him, stretching taut the already weakened skin on his back.

What disturbed me, and what it took me a while to get over, was that his skin had been unbroken before it split. When the dust settled from the code, I called my husband and said, "My patient's back split open at work today." And he asked, "Just like in *Alien*?" To which I answered, "Yeah, just like in *Alien*, except without the alien hopping out at the end."

I wasn't the only one disturbed, either. For days afterward Penny said, "Let's not have any backs split open today, OK?" What happened to Jim would be unsurprising on a surgical floor, but for most of us, even people in health care, it violates our sense of the normal, takes us into the realm of horror movies or science fiction. The bizarre ripping noise, the sudden spread of fluid just dark enough to possibly be blood, the puddle of water next to the patient, and the gaping opening itself makes one look to see Freddy Krueger of *Nightmare on Elm Street* flexing his deadly fingernails, when in reality Jim's body did this to itself. A seroma results from the healing process going awry, and even then some seromas will drain back into the body on their own. Unfortunately for Jim, his did not.

And where was Jim in all this? The patient may seem to have been forgotten in this story, but Jim showed no awareness of the activity swirling around him. His respiratory status had worsened, along with his degree of obliviousness. His daughter had been in the room with him all morning, but she had left to get some lunch before we came to move him. Off and on all morning I had heard the doctors speculating as to whether he needed to be transferred to the ICU. It isn't always clear which way someone will go. No one expected Jim's back to split open, but it was his respiratory status more than his back that got him moved to the unit. After the doctors stabilized his breathing, they would pack the wound and maybe use a wound vac, or vacuum-assisted closure, to try to get his back to heal from the bottom up. Stitching him back together in the OR would only lead to the formation of an abscess since the wound was no longer sterile. First things first, though—they needed to get him breathing OK.

ABC's—we heard it over and over again in nursing school—airway, breathing, and circulation. Is the patient's trachea open? Is the patient exchanging oxygen and carbon dioxide? Does the patient have a pulse? These are the central questions, and they should always be asked in that order. If the patient can't breathe, fixing the hole in his back will not help him survive. Much later I found out that during our code, another code was called on a room just two doors down. In that room the nurse had come in and found her patient's oxygen level at 75 percent. She said the patient was blue. Penny and I were disturbed by the grotesqueness of our situation, while next door a patient was truly close to death. If our code team had seemed blasé about what Penny and I found so startling, it might have been because their counterparts were also on the floor helping someone who needed immediate saving. In health care, breathing will always matter more than a gaping wound, especially one that isn't bleeding and isn't causing the patient any distress. "Just slap on a 4 × 4," my unit director said, referring to a common kind of simple wound dressing that measures four inches by four inches. Well, for Jim we would have needed a 4 × 10, which doesn't exist, but even in our imaginations there was no such easy fix for the patient next door who couldn't get enough oxygen.

Most of orientation was like this: a confused jumble of things I had learned in school, sometimes complementing, but often contrasting with, the reality of inpatient health care. There were orders I couldn't understand, drugs I had never heard of, phone numbers I couldn't locate, and cohorts of people whose names I was struggling to learn. I didn't know the other nurses, and the doctors were a blur of white coats, some young and obviously inexperienced—the interns and residents—and others older and wiser to the ways of oncology and hospitals—the fellows and attendings, the intensivists, surgeons, and cardiologists. There was PT and OT, HUCs, transport, IV team, pharmacists, nurses from admissions, respiratory therapists, aides whose names I needed to learn, and, oh yeah, the patients. People say it takes a year on the floor to learn the job. It's a hard year during which the learning curve is very steep, but never steeper than during those first weeks of orientation, and steepest of all when the un-expected becomes normal: when intact skin splits open and spills its contents on the floor, when a patient who had been OK suddenly cannot breathe.

At my hospital on my floor, medical oncology, orientation lasts eight weeks for new nurses. A nurse on orientation works closely with another nurse, her preceptor, gradually increasing her patient load and becoming more and more independent. I had a strong independent streak already; hence my willingness to tell Penny, a nurse with much more experience than I had, to go ahead and call the code. My challenge was figuring out what I didn't know and how I could most efficiently learn it within the confines of a system so byzantine and idiosyncratic that at mo-ments I really would have liked to bang my head on the wall in frustration, except that I never had time.

Those stories about learning the fastest allowable rate to give potassium IV, trying to remember which color tube I needed for a vancomycin trough, or discovering where to chart a patient's swollen feet (Surprise! Cardiac, not renal) are not that interesting, but they matter tremendously in terms of learning how to be a good nurse. They speak to the task aspect of nursing, the part that isn't sexy or dramatic, but comprises the nuts and bolts of patient care. Such skills are learned through repetition, taught by preceptors, and become so much a part of the fabric of being a nurse that we tend to forget there was a time when we didn't know them.

Another set of skills comprises what nursing professors like to call "critical thinking," but really is applied knowledge. A situation arises, such as a patient's back splits open. At the moment when that happened, the only trick I had in my bag was "Call the code." Now I also have "Slap on a 4 × 4" and "Call the doctor." Over time a good nurse will learn when to listen to that funny feeling in her gut and have some idea what to do about it. Should I put my patient on oxygen? Give some hydralazine IV? Suction the blood clots out of her throat? Insist that a doctor see the patient right away? Or call the code? Intuition is often looked on slightingly as the basis for any sort of practice, but for a good nurse, intuition informed by knowledge becomes priceless common sense.

Beyond all these technical skills, orientation involved learning the human side of nursing, and another experience I had early on in my eight weeks brought that lesson home. This patient, Mildred, came from rural Pennsylvania and reminded me, with her rough country speech and her understated manner, of people I knew growing up in southern Missouri. I got her as an admission, one of my first as a new nurse, and I knew only that she had just been diagnosed with leukemia.

"New leuks," the experienced nurses called them, but since

everything was new to me, the phrase didn't have much meaning. I'd had classes that focused on adult and pediatric oncology, but translating that course material, which for the adults primarily dealt with solid tumor cancers, was very hard when confronted with a real person full of questions and terrified for her life. At that point I knew nothing about chemo and only hoped Mildred wouldn't ask me anything about it. I had very little understanding of the kind of "line" she would need for IV access, and I was just learning that for some chemo regimens, patients need tests prior to getting the drugs to make sure that their heart, or liver, or kidneys can tolerate the medications. It's embarrassing to admit, but I knew almost nothing that would help Mildred understand her diagnosis or the details of her treatment plan.

I'd done the initial settling in—gave her ice water and a gown, listened to her lungs and heart, checked her pulses—and now went back into her room to try as best I could to answer her questions, or at least to discover what her questions were. Her husband, Charles, was in the room, too. They were both gray haired, and their skin had that weathered but healthy cast unique to people who have spent their lives working outside. Their expressions were dour, more from habit than a reflection of mood, I thought, because I had seen that expression on the faces of my own grandparents. They wore a look of beaten resilience, a look that said life is hard, but those who persevere survive.

When I went back into Mildred's room, I expected to encounter stoicism, a taciturn forbearance honed by years of meeting trouble head on, but that wasn't what I found at all. I started our discussion with the standard intro I had learned in school: "Do you have any questions for me?" I thought they might, using a minimum number of words, offer up a question or two, and I just hoped answering them would be easy. I didn't expect much probing, but there I was wrong. Mildred and Charles did have questions, deeper harder questions than "When will my chemo start?" or even "How long will I be here?" They were questions I could answer, but they were difficult in ways I had not expected.

"What do they say about the kind of cancer she has?" Charles asked me.

I cocked my head to one side, puzzled. Before I could ask him to explain, he spoke again, this time giving a little more detail. "What do they say about how well she'll do with the treatment?" he asked me.

I looked at the two of them and trying to sound knowledgeable said, "I really can't answer that. Only the doctors can tell you that." I knew enough about newly diagnosed patients to understand that the specifics of a patient's prognosis, based on the genetic profile of her disease and what chemotherapy was available to her, could take some time to work out.

He kept asking me, though, with variations on the same question. "How will she do with treatment?" he would say, or "What do they say about the treatment?" Mildred was silent, and I grew more and more confused, until finally he said, "We're just worried that you know something bad, and you're not telling us."

Bingo! Suddenly I understood. In nursing school the questions posed by theoretical patients were always softballs, and those fictitious patients always got right to the point. I had not been taught that a patient and a patient's spouse can be so afraid of getting a truthful answer to a question that they will be unable to ask it. Their mind won't let them form the words because the answer could be intolerable. So these worried patients, husbands, and wives ask around the question, over it, under it, behind it.

Figuring out what they really want to know can be challenging, even impossible, if their fear has led them to hide the question too deeply.

In this case I was lucky that Charles could speak the truth, and I was so relieved when he finally told me what he really wanted to know that it was hard for me not to laugh to release my own growing anxiety. I gave him the same answer I had given before, but I elaborated more to make it clear that I was not holding out on them. "I *really* cannot answer that," I said. "It really is only the doctors who can tell you that, and my understanding is at this time they do not know the answer themselves." I paused. Mildred and Charles were listening very carefully. "I am not keeping information from you," I directly told them. "I do not know more about Mildred's cancer than you do."

They both looked at me, and I could tell Charles felt relieved. He sat up a little straighter in the chair, and his shoulders became less hunched. He needed to know that we all had the same set of facts, and he also wanted to make sure that no one was keeping information from him.

Then Mildred spoke. As Charles had asked me the same question over and over again, she had nodded emphatically when I told him that I really could not give him an answer. I think she cared about the answer, but she had her own pressing question. "What's it going to be like?" she asked me. And this time I understood the question: she wanted to know what her treatment would demand of her personally.

From what I had seen of Mildred already, and from what I knew about her background, I thought she would want an honest answer, so I gathered myself together mentally. That particular question is very hard to answer truthfully, but I gave it my best shot. I looked at her and said, "You're in for a rough ride." I explained that the initial chemo would mean an extended hospital stay, that sometimes the first round of treatment didn't work, and when that happened, we often began a second round of treatment right away. I talked about mouth sores and nausea and told her we had medicines to help keep those side effects under control. She listened well, but I'm not sure the particulars of my answer mattered nearly as much as my saying, "You're in for a rough ride."

Later, though, I agonized over whether I had said the right thing. It seemed so harsh, but to pretend that the treatment was physically easy to take and always worked would have been a lie. To hedge, to say, "Well, we'll have to see," would have been more true, but still incomplete. Mildred was tough; she knew the score, she knew what it meant to have leukemia. She wanted to know what she was in for. What would she have to endure to save her own life?

It turns out I was more right about the course of Mildred's treatment than I ever wanted to be. The doctors tried several different chemo regimens, but Mildred never "cleared"—the leukemia never left her bone marrow. She developed intractable diarrhea and finally got too weak to go outside and smoke, a habit she'd kept up, even in the dead of winter, all through her hospitalization. Five months after her initial admission, after short visits home and repeat readmissions for fever, nausea, vomiting, and, of course, the disease itself, she finally went home for the last time, on hospice.

Florence Nightingale called nursing "one of the Fine Arts" and described it in terms of artistic production: "Nursing is an art: and, if it is to be made an art, it requires an exclusive devotion, as hard a preparation, as any painter's or sculptor's work." These two forms of visual art are an interesting choice. She could have compared nursing to farming, religious service, the care of animals, or even medicine, but she chose painting and sculpture, art forms that require inspiration and vision combined with a high degree of technical ability.

In speaking of devotion, Nightingale might have been echoing contemporary proscriptions that said nurses needed to be spinsters and ugly, but taken in context of her comments about painting and sculpture, I prefer a more human interpretation. To be a nurse, you have to care. You have to care about people not falling off chairs and hurting themselves, and you have to care about people's desire to know the truth about their own disease. At times this caring will ask so much of you that being devoted to the job is the only thing that will enable you to keep doing it.

The skill set you need as a nurse will stretch from hands— literally used to hold someone in place—to heart—the patience to listen for the question behind the question, the courage to give an honest answer. It's called nursing practice because it can make physical, mental, and emotional demands that no one feels prepared for when they first come onto their floor. The beginning has its share of oh-my-God moments, as in "Oh my God, his back split open," and quite a few "What do I say now?" moments as well. Each patient comes to us as a blank canvas or a solid block of stone, and at first we will make only the simplest of brushstrokes, the most obvious chisels.

At some point, though, sooner than any of us would wish, our artistic mettle will be tested. A patient will be in great distress, and it will be the nurse's job to help her. Colors and brush ends will fly, and metal will strike stone, chipping off chunks more or less artfully. In the end the product will not resemble the *Mona Lisa* or Rodin's *The Thinker*, but a real person in less distress, physical or emotional, than she was before the nurse came into the room. My masterpieces are all internal: ease given to a suffering human heart.

THREE

First Death

Almost every job has its own initiations and rites of passage. For small business owners, it's the first dollar earned; for lawyers, the first case won. Journalists take pride in their first front-page story, and teachers remember their very first classroom. In nursing, especially in oncology nursing, the first death is a professional rite of passage. This puts nurses in strange company: with doctors, of course, funeral home owners, police officers, soldiers, and assassins. My first death came when I was still on orientation, further along than when my patient's back split open, but still being supervised by Penny. Unlike some professional milestones, this one came unanticipated and unsought: for who, even in health care, goes to work ready for someone to die?

The patient, Mary, who was my first death had no treatment options left. A CT (computed tomography) scan had shown that her lungs were completely taken over with disease, and her medical team had refused to put her on a ventilator because they weren't sure she would ever be able to come off it. She was wearing a face mask that provided the same amount of breathing support as a ventilator, just without needing to have a tube stuck down her throat. The only reason she was allowed to stay on our floor with that level of respiratory difficulty was because she had a DNR/DNI order: do not resuscitate, do not intubate. A patient this unstable would normally be in the ICU.

In addition to getting help with her breathing from the mask, Mary was also on a PCA, a pump that gives a continuous dose of morphine. PCA stands for patient-controlled analgesia, but Mary was not conscious enough to supply herself with pain medication as needed, so she was getting it just like any other IV drug, with the possibility that we could bolus her: or give her added doses as needed for what we call "breakthrough pain."

As an example of the kind of coincidence that becomes routine in the hospital, I had admitted Mary a couple of weeks earlier. Emily was my preceptor that day, and she and I were told we were getting an admission from the outpatient part of the hospital's oncology division that is located right across the street. Admissions from the outpatient center were fairly common: patients who had fevers were sent over, patients who needed monitoring during transfusions were sent over, and patients who just weren't doing that well were sent over. This patient fell into the latter category, although the details we got in report were vague. There was something about "having a little trouble breathing" and not much else.

Mary arrived on the floor with her husband, Al. She was in her late sixties, wore rather large glasses, had a pinched nose, and wore her short white hair swept up and back from her forehead. Her husband was a hulking guy who seemed almost too big for whatever room he was in. He was also white-haired, but had just a thin halo of it wrapped around his mostly bald head.

When Mary got to the floor, Emily and I saw right away that she was doing much worse than the admission report had led us to believe. She was breathing with the help of oxygen she had brought from home, but she was visibly struggling for breath, and her heart rate was through the roof: 160s to 170s. This is an emergency. At that level of tachycardia (rapid heart rate), the heart's ability to pump blood is seriously compromised. The heart itself is beating so quickly that the atria, the upper chambers of the heart, do not have time to fill up with enough blood for the ventricles to pump sufficient volume out to the rest of the body. One symptom of this can be shortness of breath, since our blood carries oxygen to the rest of our body, but usually it just makes people feel "really bad" or "strange." As vague as those descriptions sound, feeling "really bad" is usually a very bad sign. Whenever a patient feels "weird," "terrible," or "bad," it typically means that something in their body is seriously not working, even though they can't say exactly what.

Seeing how fragile Mary was, the floor sprang into action. She needed to be switched over from her home oxygen to our oxygen, which was built into the wall and could give her a bigger volume of O_2 than her home unit supplied. I went in search of tubing and the attachments we needed, hoping I could remember where they were. Someone called the intensivist on duty and a different nurse brought the crash cart into the room. This was a code situation; Mary was unstable enough that Emily or I could have called a code. But our charge nurse knew that Mary did not want to go to the unit—the ICU—so we were hoping to stabilize her on the floor.

Feeling all thumbs, I got the oxygen hooked up, the intensivist arrived, and another nurse put the pads from our portable defibrillator on Mary. To me it looked like a swirl of chaotic activity, but each individual nurse knew what to do. The intensivist was, again, Dr. Sutherland. This time I noticed that he had a bit of cowboy swagger to him, and in manner, if not really in looks, he reminded me of the actor Chris Cooper. However, like any good action hero, Dr. Sutherland is imperturbable in situations where the rest of us feel panicked. He asked a few questions, checked the heart monitor, looked at the patient, and said, just barely moving his lips, "OK, let's give her some adenosine."

In nursing school, adenosine is one of those drugs that come with their own aura. It can switch patients out of particular kinds of dangerous cardiac arrhythmias, but sometimes it briefly stops the patient's heart to accomplish that switch. The trick to giving it is to give it fast, and that means as a push, an IV medication that goes straight into a vein, and for adenosine, with as much force applied as possible. Adenosine for this purpose is stocked on every crash cart. It comes in a special box in a prepackaged delivery system so that in an emergency no one has to fumble around with a needle and syringe, drawing the medication out of a vial before injecting it.

I stayed in the room to watch Emily give the adenosine, telling myself that it was a learning experience, but also admitting that I was curious to see this powerful and scary drug given to a patient. Emily's hands were shaking as she got the adenosine out of its box. I asked if I could help, and she snapped out, "No." Giving this drug at this time to this unstable patient was *her* responsibility—there was nothing *I* could do.

Emily gave the dose, but before she did, Dr. Sutherland, whose usual bedside manner ranges from businesslike to brusque, told Mary, "This is going to make you feel bad." He was straightforward but sympathetic, and that made an impression on me: giving adenosine called up a more human response from him; he felt a need to prepare Mary for what was to come. When Mary's reaction did come, it was otherworldly. Her whole body shook, and she looked terrified, adrift—unmoored from any sense of being intact and whole, as if she feared she was going to shake apart. Her face was a study in panic . . . and then it passed. She converted, as we say, to normal sinus rhythm, a regular heart rhythm at a regular rate. She relaxed, she breathed easier, and I left the room. My curiosity was gratified, and I had had my learning experience; right now Emily's and my other patients had needs, too.

Before I left the room, while Emily was giving the adenosine, I happened to glance into the back corner of the room. Al sat alone in a chair, as far removed from the action as he possibly could be. If his wife was a study in panic, he was a study in misery. The look of pain on his face as he watched Mary suffer through our life-saving ministrations eloquently bespoke the bond between the two of them. You must love someone very much to ache for her in such a visible and isolated way, I thought, and anyone feeling that much love will hate having to sit and do nothing while his wife almost dies in front of him. I looked at him. In the rush of activity he had been forgotten. He looked at me, and I smiled at him. His reaction to that smile was even briefer than Mary's reaction to the adenosine, but for just a second he looked as if he was no longer alone. His misery overwhelmed him again, but he seemed buoyed, at least a little bit, and I felt that despite being all thumbs, I had something to contribute after all.

When I walked into Mary's room two weeks later at the start of day shift, I did not recognize her. The breathing mask covered half her face, but more than that, the imminence of her own death had taken away some essential part of who she was. This happens: a person's body reaches a critical point beyond which she cannot be saved or even helped, and the individual's humanity recedes as her physicality takes over. Death is the final stage in that process, since in death the person's body remains, but her spirit, or soul, the force that animated her and made her who she was, is gone forever. Perhaps if our bodies vanished when we died, death would be easier; part of the puzzlement of death is that the body stays, but the person we knew and loved will never come back.

When I walked into the room, Mary was conscious but unaware. Despite the large doses of morphine she was receiving, she was agitated and struggled to remove the breathing mask. Her struggles with the mask were some kind of instinctive response, to the discomfort of the mask itself, to her own difficulty breathing, or both. She was experiencing what we call "air hunger," a situation in which a patient cannot get enough oxygen circulating regardless of how much we pump in. Patients with air hunger feel terror at a primary level. They are, literally, suffocating. Penny had come into the room and saw Mary's arms flailing. "Your patient is in distress," she said. "What are you going to do?" Morphine helps with air hunger, but obviously it wasn't giving the patient enough relief. I called the doctor and got an order for Haldol, an antipsychotic, which calmed her.

Her family had gone home the night before, but around ten o'clock that morning one of her sisters called and said they were planning to come in later in the day, but could come in earlier if they were needed. I told her the sooner they could get to the hospital, the better, but they did not need to rush over. The family had a strong sense that Mary was close to death, and her sister also asked me to contact a priest to perform the last rites. I thought that finding a priest would be easy, but it turned out to be complicated, maybe because it was Thanksgiving weekend. I tried the hospital operator, who gave me a pager number that turned out to be the wrong number. She then gave me two more pager numbers, and I left messages with both of them. In the middle of these phone calls and pages a different relative, a daughter-in-law, called to tell me the family would show up around noon. Penny thought I should delegate priest finding to the unit secretary, that it wasn't part of a nurse's job, but I couldn't see doing that: I promised the family I would find a priest, and I wanted to keep that promise.

Finally, after another half hour had passed, a priest called me back. He had a long Nigerian name, Umanankwe, and an accent so thick I had trouble understanding him, but he was willing to come and could meet the family's schedule. We exchanged phone numbers, and I told him which hospital, which floor, and which room. I marveled at how odd it seemed to page a priest. Shouldn't priests need a more ethereal form of communication? About a half hour later a different priest returned my second page. This one I could completely understand, but he seemed to lack the seriousness of the African priest. When I told him another priest was already coming to the hospital, he almost laughed with relief. I felt shocked, then later realized how arrogant I was being. After all, who wants to leave home or work, especially on a holiday weekend, to help someone die? I was scheduled to be there, and the priest was not—surely I could allow him to be a human being, not to welcome death intruding at all possible times into his everyday life.

The priest and some of the family members showed up at almost the same time, and I introduced them to each other. The priest seemed incredibly kind, but I was worried how the family, a group of white Catholics from a blue-collar part of Pittsburgh, would accept this man with the thick accent and the very dark skin. It turns out that in this case, too, I was being simpleminded. The priest had given a sermon at this family's church, and as a group the family was friendly with the bishop the priest worked for. While the remaining family members gathered, the family discussed parish matters with the priest as if they had all known each other for years.

Finally, the entire extended family gathered, including the patient's husband. Seeing him, a light went on in my head. I remembered the whole episode now, the lonely, tortured husband sitting in the chair watching his wife as her heart raced and she struggled to breathe. Had that been just two weeks ago? I looked at him, and he seemed to know me. But was it me, I wondered, or the memory of a smile that cut through his fear? With everyone present, the priest could perform the last rites, and he did, simply and beautifully. While Father Umanankwe performed the ceremony, Mary's room became a holy place, a haven without earthly concerns. Then he left, and the hard work of dying and grieving began in earnest.

The family had questions for me about Mary's occasional groans and seemingly purposive movements. Was she really conscious, they wondered. Why did it seem as if she were gesturing toward one or another of them, trying to communicate? I realized they wanted to know whether she herself was still in her body. I went into the room and first tried to talk to Mary. I called her name, then briefly ran my knuckles hard over her sternum. This is called a sternal rub, and it's usually the first thing emergency responders do to see if someone is conscious. If you do it hard, it really hurts and can leave patients bruised. Mary did not pause in her motions, cry out, or in any way respond to what I had just done.

I looked at the family and explained what I could in as simple terms as I could find: "When people are near death, they can have reflex movements like this that make it seem as if they're aware, but really their gestures are reactive, just impulses from the nervous system." Never in my life have I had such a rapt audience, not when speaking to my own children and certainly not when I was a professor. The family members listened so carefully to what I said that every word felt painful and faraway, as if I were speaking through water, as if, and this was true, the end of someone's life hung in the balance. I spoke slowly and looked around the room. I talked about the respiratory mask and the level of support it was giving her, explaining that without it she would not be able to breathe. In so many words I said, "This is the end of her life; this is death."

By this time it was around 1:00 P.M., and we were all waiting for the attending, the senior doctor on the case, to come and describe the options to the family. Mary could live for a while longer with the mask on, and the family needed to hear that from the attending so that they could make their own choice about what to do. I had spoken with the resident several times that morning to arrange a meeting after the last rites were over. The meeting finally took place, but the attending punted. "We have to wait for the family to come to us," she said. Penny agreed. "That's what you do," she said. "You wait for the family to come to you in their own time." I couldn't believe what I was hearing. That's bullshit, I thought. We put that breathing mask on her; we have a responsibility to give them permission to take it off. The doctor's position seemed pathetic and cowardly. Why should we wait for them to come to us? Wasn't it my job to help people face this? What kind of a nurse would I be if I saw this family and this patient suffering and didn't offer any real help?

Over the next couple of hours I had a serial conversation involving one of Mary's sons, the daughter-in-law, and two family friends who were also there for the vigil. I found myself in a difficult situation because I wasn't sure what the family wanted. I didn't think it was fair to wait for them to "come to us," but I also wasn't sure if they wanted me to come to them. Mary's two sons, her husband, a cousin, and the two friends had gathered in our patient kitchen when the two friends indicated that they wanted to speak to me. We stood in the doorway of the kitchen, a little apart from everyone else, and they talked to me in soft voices: "We thought the doctor would give some direction, would say it was time." They were earnest, in pain, and puzzled. "That's what I thought would happen, too," I said. And then I knew that the family was ready for Mary to die, but amid our technology, routines, and hierarchies, they did not know how to make that happen.

The resident had understood the situation better than the attending, and before she left for the day she gave me instructions on how to remove the breathing mask. With a patient as starved for oxygen as Mary was, removing her respiratory support would cause immediate and dramatic air hunger. Even if she could not feel pain, she would struggle for breath, physically grasp for it, so strong is our instinct to breathe. It is painful beyond belief to watch a loved one literally dying for air: it taps into a primal fear and stimulates a profound feeling of powerlessness. However, air hunger can be prevented with morphine, and the resident gave me orders for administering a bolus dose of morphine before I took the mask off, then every fifteen minutes afterward until the patient was CTB: ceased to breathe. She had no idea how many doses of morphine it might take before Mary died.

I went back to the family in the kitchen, and Al started talking to me about his frustrations with Mary's care. He was very angry at her primary oncologist, and I had been on the floor for such a short time that I didn't even know who it was. Al said this doctor had delayed Mary's chemo, and he wasn't sure why. "If that could have saved her life," he said, "that doctor's going to be hearing from me." Al was a big man, and right at that moment he looked menacing. His anger was deep and real. He wanted to confront that doctor, to hurt him physically if delaying Mary's chemo had cost her her life. Then he looked at me. "I'm just not ready to let her go," he said. "I'm just not ready." I stood in the kitchen and listened to Al. I said nothing I just listened; then my phone rang, and I had to go. But fifteen minutes later one of the sons came to the nurses' station and asked for me. "We're ready," he said simply. His eyes filled with tears. "I just can't see her suffer like this anymore; I just can't watch it." I went into the room and asked if everyone was ready, and the seven people in the room, including Al, looked at me and nodded.

Since I had never taken a patient off life support, I needed some guidance. One of our nurses, Beth, had previously worked on an oncology floor in a Chicago hospital where they had a lot of hospice patients. She came to me and explained the process: "You'll bolus her and then take the mask off."

"Do I bolus her with a shot?"

"No, you just bolus her from the PCA," Beth said. "That's called a loading dose." Beth is at least ten years younger than I am, but already she seemed so wise. She looked at me, "I can help you if you need it." I nodded. I thought I understood what to do, but I was very glad to have her offer of help.

I went into the room and explained the process to the family. I told them the doses of morphine I would give and when and why. I spoke slowly and clearly and made sure there was no dissension in the room, no shaking of the head, no expressions of silent outrage. I nodded back at the family and went to get the key I needed for the PCA. As I left the room, Mary's elderly mother, who had been at the bedside all afternoon, grabbed my arm and said, "Thank you for caring so much." I looked at her, stunned at her generosity—her daughter was dying in front of her eyes, yet she had time to think about me.

I came back in with the keys, and Beth hovered in the doorway. Narcotics are so tightly controlled in hospitals that an actual key is required to change the dose on a PCA pump. It's a circular key, like the ones that come with U-shaped bike locks. I put in the key, turned the key to activate the pump, and pressed "loading dose" on the on-screen menu. I programmed in the dose, then stood there while the small screen on the pump slowly counted off each 0.1 milligram. And then—there's no other way to say it—I stepped near the bed, right beside Mary, and took off the mask.

The patient was calm, but the breathing machine erupted. We had triggered an alarm, and you couldn't stop it even by turning off the machine. I suppose that, in theory, such an alarm is a good idea, but we hardly ever used these machines on the floor, so no one knew how to shut it up. The family stood there watching Mary while the machine gave off its high-pitched pulse. Finally,

Beth took the entire machine, trailing wires and tubes, out of the room and down the hall, beeping all the way.

I still had three other patients, and one of them, Sal, and his wife had spent the day waiting, more and more impatiently, for his discharge. First there had been an insurance hang-up, then a problem with needing the attending, rather than the resident or intern, to sign some of his prescriptions. I had fifteen minutes until I needed to give the next bolus of morphine to Mary, and I used that time to prepare Sal's discharge instructions. I checked the list of medications, made sure his follow-up appointments were written down, and reviewed the care instructions for the tubes he had that were draining a very painful abscess. Another nurse always has to review and sign off on discharge instructions to make sure the list of discharge medications is correct. I got that done and looked at my watch: fifteen minutes had passed; time for the next bolus.

From the flurry of preparing discharge paperwork I walked back into Mary's quiet room. She was slightly agitated, jerking her hands in a way that looked purposeful and desperate, even though her gestures were random. I asked the family, "It's been fifteen minutes. Do you want me to give her another dose of morphine?" I sensed hesitation, but then the son who had called me into the room just fifteen minutes before looked at Mary struggling and nodded at me: "Yes." I took the special round key out again, gave the bolus, quickly—and I hoped empathically—glanced around the room, and then left, picking up the discharge papers and hurrying back to Sal's room as soon as I was out of Mary's room. By this point Sal's wife was extremely frustrated with the discharge process, and, truth be told, with me. I wanted so much to say, "I'm really sorry, but someone is dying three doors down, and I've been busy with her," but I didn't say that, and I wouldn't. It's tacky and thoughtless to tell one cancer patient that another cancer patient is dying, but it would also have been a cheap way for me to assuage my own guilt at not being able to be in two places at once. Theresa, I told myself, you weren't up to par today for them. Live with it. I went through the discharge instructions with them, and just as I had finished, my phone rang—it was another nurse on the floor: "The family wants you. I think she's passed."

Death. I had a cadaver lab when I took anatomy, so I had been intimate with dead bodies, but never yet as a nurse had I been required to go into a room and confirm that a patient was indeed dead. It is the oddest feeling in the world to put a stethoscope on a patient's chest and not hear a heartbeat, to listen to the lungs for breath sounds and hear . . . nothing at all. I looked at the family and nodded. I couldn't get my words out. I, who hate euphemisms, could not bring myself to say, "She's dead."

"She's gone," I finally told them. Al had once again been a silent observer of his wife's suffering, but at that moment he burst into tears. I left the room. I didn't want to intrude on their privacy, and since a nurse cannot officially pronounce a death, I needed to get a doctor to confirm that Mary had died. The system requires that official pronouncement, but I also needed it for myself. What if I had just missed the heartbeat? I wondered. What if she started to breathe again? What if she had reached up and grabbed my arm, like something out of a horror movie? But that's the thing about death: it is completely final. Dead people do not breathe again, grab arms, sit up, speak, and that's what makes death so profoundly sad. When you see a corpse, you learn what it means for something to really end.

The on-call resident came and pronounced the patient, and the paperwork came next. In Pennsylvania the death certificate is a one-page form, and there are rigid rules for what the doctor fills out and what must be left blank, although the form itself in no way indicates these directions. The form also can only be completed in *black* pen. Forms in blue pen will . . . what? . . . be thrown away? And what about the forms with incorrect spaces filled in? Are those people less dead or perhaps not even dead? If only we could save lives by simply filling out a form in the wrong color ink or putting a stray mark in a sacrosanct space. It occurred to me that the only sure things in life are not death and taxes, but rather death and the idiocy of bureaucrats.

However, the paperwork got done, the family came to understand that an autopsy would be done only at their request, but was not required, and a funeral home was contacted once they knew the patient could be spared a trip to the morgue. The funeral home would come to take the body away, and I spoke with the director there a couple of times about when to show up. Funeral home workers don't like to come to the hospital when family members are still present because it creates awkwardness. The family finished saying good-bye to Mary, then said their good-byes to me in the hallway. Mary's mother caught my attention again: "Thank you for everything. I look at you, and I just feel stronger," she said. I looked back at her. I had no idea what she had seen in me, but I was glad she had seen whatever it was because it helped her. The son said good-bye also, repeating words that all of us in oncology get used to hearing but still feel moved by: "I don't know how you do this. You all are angels." At that time I still wasn't sure about hugging patients and family members, so I hugged only a couple of them and waved good-bye to all the rest. I looked again at Al, literally bent with his grief. How lucky any of us are in life to know such love.

Two days later I opted out of a family ice-skating trip and instead took the dog for a long walk to try to clear my head. A cold November rain was falling, and the sidewalks were slippery. At first I walked slowly and carefully, but as I began to think more and more about my first death, I picked up the pace. Lost in thought, I crossed the street and headed down a stretch of sidewalk that I didn't realize was uneven and covered with wet leaves. Suddenly, my feet flew out from under me, and I saw myself, as if I stood outside my own body, falling in slow motion. I screamed as I fell and involuntarily yelled out "Help!" when I landed. But it was a rainy Sunday night in the Squirrel Hill neighborhood of Pittsburgh, and no one was going to hear me. Our dog, a normally very excitable shepherd mix, stood completely still, waiting. I could not stand up. Something had happened to my left knee when I slipped, and as I fell, I, who know nothing about orthopedics or knees, instinctively felt, just as Mary had felt when we first admitted her, this was really bad.

Then another thought came to me as I lay on my back in the rain unable to stand: a memorial to Mary and her loving family, to everything they had suffered through that last day. The thought came unbidden, but it comforted me all the same, and it would keep comforting me, through eight weeks on crutches and three months of physical therapy, through my time off the floor doing paperwork instead of nursing. Thank God, I thought. At least I'm not dead.

FOUR

Benched

I lay there on the sidewalk in the dark and the rain, unable to stand, when I suddenly felt that I could stand if I tried hard enough. My right leg was fine, so I rolled off the sidewalk onto my right side and pushed up off my right foot onto my shin. The dog, Patty, had remained still for longer than I had ever seen her. How did she know? I wondered. And what did she know? Using my right hand to push off the ground, I stood and lurched over to one of the cars parked on the street. I planned to use the car as a crutch for my left side, and once I got there I was stable enough on my feet. My left leg did not hurt, but it did feel odd, useless.

I put my right hand in my coat pocket and remembered I had brought my cell phone with me. This phone, I realized, was a lifeline—the difference between getting help where I was and crawling in the rain toward the closest house. I called my husband at the ice-skating rink, but I only got his voice mail. Next I decided to call friends, starting with the geographically closest friend first. As a last resort I knew I could call an ambulance, a solution I found deeply embarrassing: as a nurse I take care of people who arrive in ambulances; I shouldn't need one for myself. Then I suddenly realized that when I fell, my role had suddenly changed from nurse to patient. I was still learning how to manage other people's health problems, and now I had a big one of my own.

I reached my friend Judith, who came and got me right away. "Home or the ED?" she asked me. The ED? I thought. Why? I'll just take some Motrin, use an ice pack, and be better tomorrow. I wasn't scheduled to work until the day after. I'm a nurse, I thought. Nurses don't get sick and go to the ED. "Home," I told her. I figured my one day off would give me plenty of time to recover. And then it hit: as suddenly as I had fallen, the pain came. In response to a severe injury, the body releases endorphins, the chemicals responsible for a runner's high. Endorphins mask pain and decrease anxiety, adaptations that allow humans to keep moving despite being hurt. In evolutionary terms, this release of endorphins means that if you sprain your ankle escaping from a saber-toothed tiger, your endorphins could anesthetize you enough to allow you to keep running until you are safe in your cave. You might permanently screw up your ankle, but you would save your own life.

Judith's car became my cave, and halfway through the five-minute drive from where I fell to my house, the endorphins wore off, and I felt the pain. I had endured labor without narcotics; my dad once accidentally slammed a car door on my thumb; I'd had cuts, bruises, and sore throats; I'd pulled a hamstring, but I had never felt pain of this intensity and depth. It shocked me, like being dropped naked into a tub of freezing water. "This— hurts—like—hell," I said through gritted teeth. And then I knew. I would not take two Motrin and be well in the morning, or even the next morning, or the morning after that. Something in my knee had been torn up, wrenched, and I would need help, maybe lots of help, to get well.

By this time I had spoken to Arthur, and he arrived home soon with the kids. Judith took the kids inside, and Arthur helped me lurch, in the rain, from the passenger seat of Judith's car to the passenger seat of our van. Then he drove me to the emergency department of my own hospital. The security guard brought me a wheelchair, and Arthur wheeled me into the small room that functions as a reception and triage area, as well as an entrance to the rest of the hospital. No one else was waiting to check in with the receptionist, so I got in to see the triage nurse right away. The triage nurse decides how sick or hurt you are and how quickly you need to be seen. She asked me all the questions I expected: "How did this happen?" "What hurts right now?" and with more concern, "Did you hit your head?" and "Did you lose consciousness?" Then she asked me the question I had been anticipating during the short drive to the hospital, "On a scale of 1 to 10, how would you describe your level of pain?" I smiled when she asked me this. All day at work I would ask patients this same question—using a "numeric pain scale" is part of a health care initiative to make sure that patients' pain is appropriately treated—but I found it almost impossible to answer it myself. "Yeah," I said, "I knew you were going to ask me that." I thought about it, "Um . . . 7," I decided. When patients tell me their pain is 7 out of 10, I say, "That's a lot of pain." Was I experiencing that same level of pain? The pain someone feels when a tumor pushes against their gut? Or when cancer wastes their pancreas or attacks their bones? "Seven," I said, realizing for the first time how little absolute meaning the pain numbers really have.

Because I'm a nurse, I know that whatever you can use to your advantage to get good care in a hospital should be used, so I told everyone in the ED that I was a nurse at that hospital. This information spread quickly, and as I got transferred from the triage nurse to the ED nurse to the X-ray tech, and back to the ED nurse, I saw each person pass that information along to the next. Sure, we take care of our own, but much more than that, no nurse on staff on a hospital floor wants a nurse from another floor to act as a mole, spying out their deficiencies and returning to her own floor to announce triumphantly that the nursing in the ED—or wherever—isn't very good: that they left me sitting in a hallway for hours in pain, that they misdiagnosed me, that they treated me like a "civilian" instead of someone in the know. At this point I had been a nurse for less than three months, but in the hospital at least I already had some idea of how to work the system. Act as nice as possible, I thought, but when necessary be a pain in the ass.

The hospital itself, and especially the ED, also has unspoken expectations of behavior, and I sensed those immediately when the ED nurse came out to the hallway to get me. He took one look at my wet, dirty clothes as I slumped in the wheelchair and said, "What? You've been playing in the mud?" He was a young guy, late twenties maybe, slim, with a Russian-sounding last name that I can't remember and a very cute accent. Could be worse, I thought, realizing that all that was required of me at this point was to play along. He put me in a room and brought me a hospital gown and some blankets. By this point my husband had returned from parking our van and filling out insurance forms, and he got me out of my muddy winter coat and my muddy jeans and into a clean dry gown. Then my Russian nurse came back in, bringing me the biggest ibuprofen pill I have ever seen in my life. The whole time he tut-tutted and clucked over me, listening to my story, saying, "See, it's not so bad," in that cute accent, until

I said that initially when I fell I couldn't stand up. He stopped in his patter. "You couldn't stand up?" he asked. "No," I answered. He reappraised me. He inclined his head to the side, and I imagined I saw the gears turning in his head as they would be in mine. "But then you were able to stand up, right?" he asked. I nodded yes, and he nodded curtly back. Not being able to stand up—a terrifying five minutes as I lay on my back in the rain—was ultimately irrelevant since eventually I did stand.

The ED doctor came in once the X-rays had been done. He was young, probably a resident, thin, not much taller than I am, with dark curly hair. He wore black scrubs, and his last name began with an *X*. Who wears black scrubs? I thought, and, Am I really being seen by Dr. X? Then Dr. X began examining me, and I got it: the black scrubs announced, "I am not a touchy-feely, hand-holding doctor; I have attitude, and I drink way too much coffee. I am really hard core." He began manipulating my kneecap, which means that he moved it around, or tried to move it around, to determine whether it was stable in the joint. The exam hurt. "Do you *have* to do that?" I asked him, having already learned that the best way to deal with doctors with attitude is to sass them right back. When he protested my protests, I said, "Hey, don't get smart with me; I'm a nurse."

I wouldn't recommend the wise-ass approach for most doctor–patient interactions, but in this case, it worked. It let Dr. X be the authority without being overbearing, and it allowed me to be a supplicant without being too passive. The diagnosis came quickly once he checked the X-ray and finished examining me. Nothing was broken; it was most likely a knee sprain, he said. Only an MRI (magnetic resonance image) would tell for sure and, "Unless you're Ben Roethlisberger"—the Pittsburgh Steelers quarterback—"there's no way you're getting an MRI on Sunday night."

"Well," I said, "actually I *am* Ben Roethlisberger."

He didn't even pause. "You're still not getting an MRI tonight," he said. Too bad, because I would not get a correct diagnosis until I got an MRI, and I wouldn't get an MRI for another month. Being a nurse did not mean the system would work any more efficiently or exactingly than usual. MRIs are expensive. My physical therapist explained that insurance companies want doctors to wait and see if physical therapy alone can solve whatever the problem is before shelling out the cash for a scan. So I didn't get an MRI, and I didn't get a correct diagnosis, but I did get a brace with so many straps and stabilizers it looked like it was modeled on an eighteenth-century corset. I also got a set of crutches and a very quick lesson on how to use them from my Russian nurse. Dr. X told me I couldn't work for a week. "A *week*?" I said. He looked at me. "Hey, what did you expect?" he responded. "You're not eight years old anymore; things take time to heal."

Before I left the ED, I got a prescription for drugs for pain, and I insisted on one more thing: pants. In asking for pants, I was applying another principle I had learned about inpatient health care: ask for what you really need. Hospital gowns are notorious for being breezy and impractical, but some hospitals have pants as well as gowns, and I knew that my hospital did. The pants are one size fits all, very thin, and tie at the waist. However, the jeans I had fallen in, and taken off when I got to the ED, were covered with mud. It felt like adding insult to injury to put them back on. As soon as my discharge seemed imminent, I started with my Russian nurse: "Could I have some pants?" "I'm not sure we've got pants," he said in his cute accent. "We have pants on my floor," I said. And so it went; each time he came in the room—with my brace, my crutches, my prescription—I asked for pants. I asked enough times that it became clear I was not going to leave the hospital without hospital pants, so he found me some. If I couldn't get an MRI that night, at least I could limp out of the hospital in clean, dry pants.

On my way out of the ED, clumsy on my crutches and stiff with the knee brace, I caught Dr. X's attention one last time. "You look like a gimp," he said, and before I could stop myself, I blurted out, "You are such a jerk." He stopped for a minute, and I realized I had hurt him. "I was only kidding," he said. "No, I know," I answered, genuinely sorry that I had broken the rules he and I had set for each other, the same rules that govern most staff interactions in the hospital.

Teasing is the common medium of exchange among health care workers. In the intense world of the hospital, a staff member who is not teased is probably not liked, or worse, not accepted. When Dr. X told me I looked like a gimp, he meant it affectionately—I am certain of that. But at that point I no longer wanted to be a nurse with attitude and sass; I wanted to be a patient, to be needy. I wanted the doctor to tell me solemnly that I would be OK, to pat my shoulder, to wish me well. I wanted Marcus Welby or Dr. McDreamy—a made-for-TV doctor—instead of Dr. X in his black scrubs with his caustic sense of humor. But no TV doctors were on shift in the ED that night. I looked at Dr. X and smiled. "Thank you," I told him, hoping he knew I meant it.

The next few days were a blur of pain, pain medication, ice, and helplessness. I have helped many patients settle themselves down on a toilet seat that sits by the bed, what we call a bedside commode. Sometimes I feel happy to help them, and sometimes I'm annoyed that no nurse's aides are available because I'm busy and don't really have time, but I have never thought about how they feel. Loss of dignity is taken for granted in the hospital, and patients are not usually allowed the luxury of mourning their lost privacy or self-reliance. The morning after I hurt myself, I discovered how awful it is not to be able to go to the bathroom on your own. Between the crutches, the knee brace, the pain, and the swelling around my knee, I could not safely lower myself onto the toilet. Arthur helped me, and he did it with grace and compassion, but needing his help for such a basic life activity was a revelation. And it wasn't just going to the bathroom that had become impossible—almost all mundane tasks, what we call activities of daily living, were now beyond my abilities. Because I couldn't stand or bend my left knee, I couldn't put on socks and underwear, much less dress myself. I could not get myself a glass of water because it was impossible to carry liquid in an open container while walking with crutches. Since I couldn't make it down the stairs to where our kitchen is, I was completely dependent on others for food.

A few days after I fell, I went to see an orthopedic surgeon who specialized in knee surgeries but did not necessarily know that much about knee injuries. He was a young guy, Dr. Patton, a resident, tall and friendly, but talkative and difficult to keep on track. What I most remember about my visit to him is not what he said about my knee, but the story he told about my family doctor, Steven Nordmann. Dr. Patton had done family practice rounds with Dr. Nordmann, and he told my husband and me that Dr. Nordmann greeted a patient once by energetically complimenting her on her pregnancy. The only problem: the patient was not pregnant. Dr. Patton's sage summation—"You don't make that mistake twice"—was more helpful than anything he told me about my injury.

The truth is, Dr. Patton did not have any idea what had happened to my knee. He knew that nothing was broken and that I hadn't torn my ACL, or anterior cruciate ligament—the injury de rigueur for professional athletes and high school soccer players alike. He thought I might have a meniscal tear, a small rip in the membrane between the femur (the long bone of the thigh) and the tibia (the shin bone). He hinted several times that I should maybe see a sports medicine doctor without admitting he really didn't know what was wrong with my knee. He fell back on the "knee sprain" diagnosis, told me to stop wearing the brace, stop using the crutches, go to physical therapy at least twice a week, and don't do any floor nursing for at least another two weeks. At the end of all this I burst out, "But am I going to be OK? I mean, why couldn't I stand up when I fell?"

He paused, then started talking again. "Oh, don't worry about that," he said. "Man, I've torn my ACL so many times; I've messed up both my knees, and I'm still walking and getting around OK." He thought about it for a minute. "At least for the most part." Not the response of a TV doctor I hoped for, but comforting in its own way. Speaking as an orthopedic surgeon, he had just told me that, based on his own personal experience and his experience in medicine, I would be OK. For the moment, that would do.

After Dr. Patton told me to stay home from work for at least two more weeks and to continue with physical therapy, I felt imprisoned, but the kids showed me the silver lining in my newly restricted life. The morning after I fell, I woke up at home aching and unhappy. Our son came into the bedroom and said, "I'm glad you got hurt, because it means you'll be home next weekend." Direct, honest, and helpful, my son had just given me a whole new way of looking at my injury: the kids were happy because I would be home more. I could not walk, I could not cook, I could not even come downstairs to see them off to school, but none of that mattered compared with the upside—they were very glad to have me all to themselves for however long it took for my knee to heal.

That awareness of the silver lining could not always sustain me, though. For a lot of the time I was profoundly bored. I decided to read the longest book I could think of: *War and Peace*. At fifty pages a day, the entire one-thousand-page book would take only twenty days. I did my Christmas shopping online and sent self-pitying e-mails to friends. I watched Akira Kurosawa's movie *Yojimbo* and the remake of it starring Clint Eastwood—*A Fistful of Dollars*—and marveled that the same story could be told just as successfully in the American Wild West as in feudal Japan. I watched the clock. Minutes stretched into hours, and hours eventually stretched into days. I worried that I would never walk normally again.

One month after I fell, two important things happened. First, the human resources department tried to fire me, but my unit director saved my job. Second, I finally got a correct diagnosis, which meant that I could follow the best course of treatment for healing my knee. My difficulties with HR will probably surprise people less than the story of my delayed diagnosis, but I had the opposite reaction. Because I worked for a health care organization, I thought the rules for getting disability pay and keeping my job would be clear and sensible, but they weren't. As far as the diagnosis goes, it is unsurprising to nurses and doctors how long it can sometimes take in medicine to get the right answer.

Pretty soon after I was injured I tried to figure out which if any disability benefits I was entitled to through my job and how to sign up for them. Because my hospital is part of a large health care corporation, requesting and getting disability benefits is a multistep process involving different corporate entities. An able-bodied person would find the bureaucracy annoying, but for me, alternately doped up and in pain, the process became an ordeal. The colorful brochure that I received as a new hire with pictures of smiling people and text about supporting injured employees was less than useless since it implied that help existed, without being clear about how to get it. I filled out forms, talked to representatives, gave doctors' phone numbers and my own insurance information, even though we, the doctors and I, were all part of the same organization. Through it all I tried to remind myself that fraud exists and that people who work in benefits offices have to be careful, but in the end I resented being treated like a lying malingerer. It seemed that no amount of X-rays, doctors' notes, or physical evidence of incapacity was going to convince my representative that I was anything other than a parasite working the system. "So when were you thinking about going back to work?" she would ask me. "Um, when I can walk," I answered. It confused me that a health care corporation did not understand the idea of serious injury.

When the disability paperwork finally got filled out, I became eligible for 80 percent of my base pay for as long as I was benched—which I thought was very generous—but at the same time I discovered that HR planned to terminate my employment. What? Fired? I hadn't worked at the hospital long enough for them to hold my job for me. What I found the most odd was that I could still get disability benefits even though I would no longer be employed by my hospital. I'm sure in some actuarial model these policies made financial sense, but it made no human sense to me. I left a panicked voice mail for Diana, my unit director, who found out that if I could come in for just twenty hours a week to do paperwork, HR could not fire me. People would bring work to me, she said; I would just have to sit and do it. It would not be easy physically for me to come to the floor and do paperwork, but I thought I could manage it. The next week, achy, slow, and very dependent, I started going to the floor for four hours a day and in the process saved my job.

Going back to work as a paper pusher was a nice break because it got me out of the house, but the work Diana had for me to do was deadly dull. I had never known just how much checking and double-checking of documentation—what are called "chart audits"—is required by hospitals. When nurses chart, we record how our patients are doing and what we have done for those patients during a shift. Someone then audits this charting: goes through and checks whether pain was assessed correctly, whether oxygen was prescribed at the right level, whether the form "Interdisciplinary Plan of Care" was filled out every seventy-two hours. The charting is electronic at my hospital, but the audits were done on paper, and the list of items to audit seemed infinite: when and how nurses reported lab values, whether patients' skin was properly assessed, if patients in the ICU who needed flu and pneumonia shots received them. More than once I wondered why electronic charts could not be audited electronically.

Doing the chart audits also reminded me how much I missed being with patients. The opportunity to care about and for patients is what got me into nursing. Combing through records of other people's care could not take the place of real-time interactions with real people. I enjoyed the behind-the-scenes look that doing paperwork gave me, and I appreciated that when I got back to the floor, my own charting would be improved by doing this work, but it could not come close to filling the spot that floor nursing had in my heart.

A month had now gone by, and despite the passing of time, lots of ice, and lots of physical therapy, my knee wasn't any better, so Dr. Patton's office sent me to a different doctor, an orthopedist trained in sports medicine. The sports medicine building was huge, and once I got inside, I did not feel I was in a doctor's office at all. The waiting room was a two-story atrium with solid glass walls, and the inner walls were covered with sports jerseys, sports memorabilia, and inspirational slogans. The hallway Arthur and I walked down to my appointment was lined with signed photographs from famous athletes, all thanking the resident sports medicine guru for healing whatever had been wrong with them.

Another examining room, another doctor: this one a woman, Dr. Rausch, equally as no-nonsense as all the others, rushed, not old, but not as young as Dr. X or Dr. Patton, with snowy blonde hair. She listened to my story, had me sit on the table and manipulated my knee, asked about the swelling, which was better than it had been, and then asked me to walk. "I'm slow," I told her, but she gestured me out into the hallway. I walked up the carpeted, photograph-lined hallway about ten feet forward and then ten feet back, very, very slowly. Walking back into the examining room, I finally found the made-for-TV doctor I had been looking for.

"Oh my God," Dr. Rausch said. "You must be in such pain.

You're not bending your left leg at all when you walk." I didn't know what to say anymore. Even though I was a nurse, I had no knowledge base to draw on, no real point of comparison for the pain I felt. On the floor I would insist that patients got enough medication to manage their pain, but for myself? I had just gotten used to it. But Dr. Rausch looked at me, full of fire and energy, and said, "We're going to figure this out. I'll be back at this facility in two days. Come in two days, get an MRI, and then see me again. We're going to get you fixed up." It's sad to say, but sometimes in our modern health care system what the doctor orders is finding the right doctor—the person with the right knowledge base you need who cares enough to do the thinking required to heal you, especially when you've stopped thinking yourself.

In two days I came back for the MRI, and afterward Arthur and I returned to another examining room. Dr. Rausch came in and said, "I just want to check one thing." She pressed on the inside of my left knee and asked me if it hurt. "Yes," I answered. No surprise; that was the part of my left knee that always hurt. Then she pressed on the outside of my left knee and asked me if it hurt. "Yes!" I said, completely surprised. I had no idea that part of my knee was injured. Dr. Rausch reached her hand up in the air, made a fist, and then pulled her entire arm downward, elbow first. "Yes!" she said triumphantly, almost to herself. "Let me show you the MRI," she said, reaching out a hand to help me sit up and get down from the examining table.

She pulled the images from my MRI up on the computer in the room and showed Arthur and me what was wrong. She was very excited. "I had the head of radiology look at it, too. See that?" She pointed on the MRI to what looked like my kneecap and the ligaments that keep it in place. The ligaments looked pulled, stretched like taffy. "You subluxated your patella—partially dislocated your kneecap, so those ligaments have to heal up, have to form scar tissue to get back to normal." She grabbed the mouse and clicked around until she had pulled up a different set of images. "That's the end of your femur" (the long bone that forms the human thigh, the biggest bone in the body). "When your kneecap dislocated, it hit the outside of your femur, bruising the bone." She showed us an image of the bottom of my femur—the condyle—the part that connects one bone to another. More than half of my femur was obscured. "That's blood," Dr. Rausch said. "You have a huge bone bruise." She paused. "Bone bruises can take a long time to heal. A long time." I knew this was bad. When doctors say that something can take a "long time," it means it will take so long they are not sure the patient will be able to tolerate the news.

I looked at her. "How long?"

"It could take as long as two months," she said.

Two months? I had already been off the floor for a month and knew that every day a little more nursing knowledge slipped out of my mind, lost. By the end of two months I would probably have to start completely over. And what if I wasn't better in two months? If you cannot walk, bend, squat, lift, carry, push, or pull, you cannot be a floor nurse. Without exception, you must be able-bodied to do this job. Dr. Rausch's mother had been a floor nurse, so she knew what the job involved. "You can sit and do paperwork," she said, "but floor nursing? You're not doing that." She put me back on crutches, and I got a real lesson in how to use them, including how to walk up and down stairs. She gave me a knee brace and a prescription for stepping up the physical therapy. She told me to keep taking Aleve and to use ice, ice, ice. Data had shown that keeping a knee with this particular injury non-weight bearing "for a while" greatly aided healing, and that was why I needed to use the crutches. The knee brace limited the motion of my kneecap, and physical therapy would strengthen the muscles around my kneecap, preventing me from reinjuring myself while I recovered, and keeping me strong once I was walking again. The Aleve and ice reduced inflammation and pain. The fluid around my femur needed to go back where it belonged, and only then would I recover normal range of motion in my left leg. Only then would I be fully able again.

And eventually I was; with the correct diagnosis and the correct treatment plan, I slowly healed. I weaned myself off the crutches and then incrementally increased the distance I could walk and the speed I could achieve. I went back to work as a floor nurse and made it through my shifts wearing my knee brace and icing my knee when I got home. At some point the knee brace became more irritating than helpful, so I quit wearing it. I finished the two weeks of orientation that remained and the extra week they gave me to make up for time lost. I was walking again, and at roughly the same time I was working without a preceptor. Both things felt good.

During my convalescence I spent the first few weeks in a codeine haze, and when that lifted, I discovered how very afraid I was that I would never walk again and thus that I would not be able to work as a nurse. I loved my job, and I could not do it if I could not walk. I have seen doctors in wheelchairs and on crutches, with arms in slings and legs in casts, but floor nurses can never do this. Our bodies get us from point A to point B, and a floor nurse spends her entire day moving from point to point. A crippled RN cannot work in a hospital as a nurse.

I wish I had some profound realization to pass on from my injury and my protracted recovery time, but the only moral I take away from my experience is "Don't get hurt." It sounds stupid and obvious, even simpleminded, but it isn't. Pain hurts, disability leads to exclusion, and enforced inactivity induces desperation. I've seen these truths evident in all my patients; now I have learned them for myself. We all only have one body. Take care of yourself, use caution, and whatever you do, if at all possible, don't get hurt.

FIVE

A Day on the Floor

Every day on the floor is different; here is what happened on an average day. Four patients is a standard assignment for day shift on my floor, and this day my four patients were Peter, in his midtwenties; Marlo, a sixty-three-year-old African American; Tom, in his late thirties; and Dorothy, an eccentric fifty-year-old who was always heavily perfumed. Each of them had some type of blood cancer. Peter, Marlo, and Dorothy had been admitted for treatment of their disease, meaning chemo, and Tom was in for disease complications.

The day always begins the same way: at the nurses' station I pick up the papers for the four patients who make up my assignment. I staple them together, put a blank sheet on top for notes, attach tabs to write when meds are due, and get to work. Any notes I take during the shift will go on these papers, and any important information will be written down there. Losing one's papers, for a nurse or a doctor, is a disaster. Days are usually so busy that writing things down is the only way to make sure everything gets done.

I knew Peter well but did not know any of the others. He had just finished another round of chemo and was ready to be discharged after a two-week stay. Marlo had a bone marrow biopsy scheduled for later that afternoon. Tom had pneumonia and CNS involvement with his leukemia, meaning that his disease had spread to his central nervous system. Patients with CNS involvement could be unpredictable, manifesting a range of symptoms that might include confusion, poor balance, pain or numbness in the extremities, general weakness, and difficulty breathing. He already needed oxygen and could not safely get to the bedside toilet without help. Dorothy was a scheduled admission getting follow-up chemo, or "consolidation." I thought Dorothy would be my easiest patient, except that she had chronic back pain that her lymphoma exacerbated. Despite pretty large doses of narcotics, we weren't keeping her pain under control.

I learned all these details, and more, from report, which I got from night shift. Report is the narrative account that gets passed from nurse to nurse, shift to shift, on every patient. It includes the patient's diagnosis, what chemo the patient had and when, and relevant past medical history. It should also mention recent tests and their results, if known, give an account of problems currently plaguing the patient, and provide an overall assessment. Assessment always proceeds from head to toe. Marlo's was benign, meaning she had no issues with any major body system. Tom was "neurologically intact, slightly tachy, on 3 L O_2 via nasal cannula with bilateral wheezes, no nausea, vomiting, diarrhea." Peter was "pale," but otherwise "unremarkable." Dorothy had that back pain.

In the past, report was given face to face, but with the advent of technology report can now be recorded on a voice care system. The idea is to improve efficiency, but for me something gets lost in the translation from nurse to tape to nurse. A look in the eye, a grimace, a joke, a sigh of frustration, a brief story—all convey subtleties that no tape will ever capture. The tape records facts, but knowing how a patient behaves and what makes him tick matter to good care, too. "Tom," the night nurse said, "will keep you running, you know what I mean? He'll ask you for a cup, and then he'll ask you for ice, and then he'll ask you for ginger ale." "Got it," I said. He managed his anxiety by making requests, and my job was to help him do that while keeping his demands on my time under control.

In my mind I rank my patients in terms of least potentially work-intensive to most. We learn how to prioritize this way in school, and even the most novice nurse thinks about patients in terms of a need hierarchy. First came Dorothy. Her chronic back pain would probably require additional doses of narcotics, but her chemo was scheduled for evening shift, meaning that I would be long gone by the time she got it. Marlo was second. Her bone marrow biopsy might keep me busy when it happened, but until then she would not need that much from me, and the biopsy was planned for later in the day. Next came Peter, very stable now, but needing all the administrative work that went into a discharge. His brother was going to pick him up but couldn't get to the hospital before 4:00 P.M. The timing of Peter's ride meant that I might have to stay past the end of my shift to get him out the door. I hoped that Marlo's biopsy and Peter's discharge would not end up happening at the exact same time, because she might need some TLC to get through it. My wild card here was Tom. I already knew that he was anxious. In addition to his anxiety, he was very, very sick. From report I had learned that his heart was racing ("tachy") and that he was having such a hard time breathing he needed oxygen (3 L O_2), and that we could even hear the effort breathing took ("wheezes").

I checked my watch: 7:30, not yet behind. At my med cart

computer I looked up the day's information on my patients—their vital signs, morning labs, medications, and existing orders—and wrote down most of it on my papers so that I would have it available at a glance.

It was now 7:50, and I was deciding which patient to see first when Tom's call light came on. I went into his room. "Hi, I'm Theresa," I said. "I'll be your nurse today." I worried, not for the first time, that my enthusiasm combined with my white uniform made me come across more like a cruise director than a nurse. If Tom had any misgivings about my role, he kept them to himself: "Bathroom," he said. He was skinny, bald from chemo, and very pale. I helped him sit up on the edge of the bed, and that small bit of exertion left him breathless. This was no octogenarian with heart disease, but someone younger than I am. I helped him to the bedside commode. Standing was difficult for him, pivoting to the toilet draining to an extreme. Could this be anxiety? I wondered, because extreme anxiety can leave people gasping for air. I had never seen anyone get that winded just by sitting up on the edge of the bed.

"Do you feel anxious," I asked him, "or is it hard to breathe?"

"Hardtobreathe," he said, all one word in a rush.

Several things then happened at once. Transport called, saying they were on the floor to take Dorothy to a CT scan. I hadn't yet seen Dorothy, and she needed her morning pain meds, her insulin, and Zofran, an antinausea medication, before she went. Tom would need help getting off the commode and back into bed. If transport had to wait too long, the transporter would leave without the patient, meaning I would have to start the whole process all over again and probably get an angry call from

CT asking why the patient never arrived. It was always unpleasant to get one of those calls.

"Tom," I said, "I'm going to give you a little more time," and I left his room to go into Dorothy's. "Hi, I'm Theresa; I'm your nurse today. How are you doing?" I quickly wrote my name on the whiteboard in her room.

"You know what?" I said "Transport is here to take you to CT. Let me just run and get your insulin and OxyContin and send you on your way."

The transporter called me again. There was some hang-up at the nurses' station about getting the right papers to print, which bought me a little time. I grabbed an aide, and together we helped Tom, whose need for the bathroom had been a false alarm, back into bed, still breathing hard but stable. I got Dorothy's narcotic from the locked cabinet in the locked med room, drew up her insulin, and settled her in a wheelchair, more or less when the transporter was ready for her "I'll see you when you get back, Dorothy," I called out as she headed down the hall.

I peeked in at Tom, and his breathing was better, but he wanted his shades opened in front of one window, not the other, and slanted at a particular angle. Then he had a medication question. I had already taken each of his twenty morning pills out of their individual packages in front of him, but he wanted to make sure he got the brand name of two cardiac medications and not the generic. "I'm allergic to the generic," he said, and the hospital pharmacy only carried the generic. I found it very hard to imagine that he could be allergic to the generic version of a medication, but I personally have reacted to Benadryl, which we give to patients *having* allergic reactions, so anything's possible; and if he *was* allergic and *had* gotten the generic, the result could be serious, especially with his respiratory status already compromised. "Let me check," I said.

"Let me check" is a great cover phrase since "checking" can take an unspecified amount of time. First, though, I needed to know if Tom had gotten the generics of the two drugs and was relieved to discover that he had not. The brand-name drugs weren't ordered either. Hmm. I wasn't quite sure what to make of that, but while thinking it over, I went to see Peter. He needed IV antibiotics and IV magnesium before his discharge, and I didn't want to get behind; it was already 8:30. Peter liked to sleep late, so I hooked up his 9:00 A.M. meds, checking his IV line to make sure it was working, then headed back to Tom.

"Tom, Coreg and Zocor, aren't even ordered for you," I told him. "Yeah," he said, "I know. I have my own here with me." At this moment I gave a deep internal sigh. OK, I thought, I get it: worry aloud about a nonexistent problem, then parlay that worry into a new problem. "They're in that bag," he said, gesturing toward a collection of Ziploc bags on the chair in his room. I went over and picked one up. "No, not that one. The other one," he directed. I picked up a different one. "No, the other one," he said more emphatically. I picked up a third bag, rifled through it, and found a collection of pills . . . that did not include Coreg and Zocor. "They aren't here," I said, puzzled. "But I need them," Tom said, the level of anxiety rising in his voice.

I finally found the pills in another Ziploc bag that was inside a fifth bag. However, it is bad practice to allow patients to bring their own medications to the hospital and take them at will. We need to know what drugs they are taking, and we need to know the dose. He and I went back and forth about whether he would take the pills immediately or wait until I called the doctor to find out whether taking his own pills was acceptable. I paged the intern, one of the first-year doctors, to ask him about Tom taking the pills.

While I waited for the intern to return my page, I got out morning meds for Marlo. It was getting close to nine o'clock, and I like to have seen all my patients by nine. The intern called me back and said that Tom could have the Coreg and Zocor. I went back into Tom's room and gave him the specified doses of the pills, but then he insisted that I keep the pill containers with me. I figured we had a policy on whether nurses could keep track of and dispense a patient's own medications, but I didn't know what it was, so I took the pills for the moment, telling him I would have to find out if I could keep them in his med drawer or not. I asked if he was doing OK for the moment and was very surprised when he said yes. Maybe I really was helping him.

I left Tom's room and went next door to see Marlo. Marlo was a personable middle-aged lady with a huge friendly smile and, unfortunately, a sad history of multiple relapses. The bone marrow biopsy scheduled for that afternoon would tell her whether she was disease free after her most recent round of chemo. She and I talked about different nurses on the floor and confessed that we both sometimes had trouble remembering people's names. The topic came up because she remarked that she had never had me as her nurse before, and I explained that I was fairly new to the floor. Later I would wonder if not knowing me added to her anxiety.

I gave her her morning medications, then checked her line and IV fluids. While I did my assessment—looked in her mouth, listened to her heart and lungs, checked her belly and pulses— her anxiety about the upcoming bone marrow biopsy became more and more evident. I tried to reassure her by telling her we could sedate her beforehand with Ativan, a benzodiazepine that can be given intravenously. It can be tricky bringing up anti-anxiolytic drugs because some patients get insulted when the nurse suggests they might benefit from a benzo. I tentatively mentioned Ativan, and Marlo jumped on it. "Yes," she said. "IV Ativan it is," I told her.

As I walked out the door, she made it clear that she was counting on me to be there, with the Ativan, when it came time for the biopsy. She began joking with me again about forgetting names, pretending she had already forgotten mine. "Remember ol' what's-her-name?" she said with her gentle voice, and, laughing, I countered with, "Who forgot your IV Ativan? No, Marlo, that's not going to be me." Because really, she wasn't joking, and she and I both knew that.

It was 9:30, and I had gotten through what would be, I hoped, the busiest part of my day. I was not too far behind. I still needed to do an assessment on Peter when he woke up and on Dorothy when she returned from her scan, so I quickly made notes in the computer showing that I had seen both of them, that they had "no complaints," and that I would do a full assessment later. Peter needed a few more meds, so I briefly woke him, checked out his major systems, and got him to swallow some nasty-tasting liquid medicine. He did what I asked, then fell right back to sleep.

Meanwhile, Tom's call light was going off. I sighed again internally and went into his room. "Hi, Tom, what do you need?" "Bathroom again," he said. I went to the side of the bed and helped him again to the bedside commode. He experienced the same windedness that he had earlier, but this time he went to the bathroom. I handed him toilet paper and wet wipes, and he wiped himself. Then it was time to get him back in bed, but by this point he was so winded that he needed more oxygen. I turned his oxygen supply up to 5 liters from 3, and though he was still breathing very quickly, he said he felt better. It was at most a foot of pivoting from the commode to the bed, and slowly he and I did it together. Once he was back in bed, he sat straight up, gasping for breath. I had time to wait in the room, and I wanted to see how he would do if I turned his oxygen down. His breathing slowed to a more normal rate in a couple of minutes, and I turned his oxygen to 3 liters. "Tom, you're back at 3 liters," I told him. "Does that feel OK?" He nodded and lay himself back against the pillows on the raised head of his bed. Then he asked, "What about you keeping the Coreg and the Zocor?" "Oops!" I said, raising my hands in annoyance at myself. "Let me find out right now."

Once again, several things happened at once. The team came to do rounds: the attending M.D., the fellow, the resident, the intern, and the pharm D, or doctor of pharmacology, who was also part of the team. The pharm D, Gayatri, a tall, beautiful Indian woman, already knew about the issues with the Zocor and Coreg. "Patients can't just take their own medications," she told me. "The intern doesn't really know." She frowned, making a crease in her forehead. "I have to find out about this."

Then she and I had one of those conversations that would have totally mystified me when I started on the floor, but that now I understood. "Also," she continued, "we have questions about his vanco [vancomycin] level and his tobra [tobramycin] level."

I nodded. "The vanco trough was taken last night after the vanco had started," I told her. "That was before he got transferred here. Somehow the order got put in that way."

"Oh, so that's completely useless," she said, wrinkling up her brow again. Then she returned to the tobramycin. "The tobra level is really more important anyway, because we haven't gotten an accurate level on him yet."

Tobramycin and vancomycin are both powerful antibiotics, and unfortunately the concentration at which they are most effective is very close to the concentration at which they can cause toxic side effects in the ears and kidneys. For this reason, we draw blood samples to determine the concentration or "level" of the drug in patients' bodies so that the dose can be adjusted if necessary. A trough level is drawn when the drug is at its lowest amount. A peak level is drawn soon after the drug is given, when the highest concentration of drug possible will be in the patient's blood. Gayatri explained to me that the trough should be drawn at 11:30, I should administer the drug at noon, and I should draw the peak level at 12:30. It sounded simple enough, but the difficulty would be in the timing. Nurses on my floor can only collect blood for labs if the patient has a permanent or "central" IV line. Tom did not have such a line, so his tobra level would have to be drawn by one of our phlebotomists, and even if I called the phlebotomist and told her the lab was "stat"—meaning right away—she still had to come on her own schedule, not mine.

Gayatri and I finished our conversation right when the attending M.D. came out of Tom's room. I liked this attending, Dr. Girlitz, with his dark hair and surprisingly boyish smile, but with him conversations were always fast and furious. He started talking, with his slightly accented English, as soon as he left the room. "He's using these accessory muscles," he said, running his fingers up and down his own neck. "If that doesn't stop, he's going to code."

"What?" I wanted to shout. "Do you mean that?" But instead I said, "Yes, when I helped him to the bedside commode, he became extremely tachypneic. I turned the oxygen up to 5 liters."

"What were his O_2 sats?" he asked, referring to oxygen saturation levels.

"Mid-90s the whole time."

"Where is the O_2 now?"

"I turned it back down to 3." Trying to be helpful, I said, "And it really is shortness of breath; it's not anxiety."

"Yes, yes," he said, impatient now, "because he's using these accessory muscles." He ran his fingers up and down his neck one more time. And then the team moved on.

Now I was concerned. I found a nurse on the floor with more experience than I had. "Dr. Girlitz just said that if Tom's breathing doesn't improve, he's going to code." I swallowed. "I love it when doctors say things like that," I told her, trying to relieve my own anxiety with sarcasm. "It really gives a shine to the whole day."

This particular nurse, Karen, had very expressive eyebrows, and right now she raised them. "Maybe you could just send him to the unit now instead of waiting for that to happen," she said.

I looked at her for a second, and then the wisdom of what she had suggested sank in. "Great idea!" I told her. "I'll work on that." I'd learned from experience that it was always better for the patient not to call a condition; whenever possible, to act before they became critical.

Time passed. Gayatri told me that Tom could have his Coreg and Zocor, the heart medications, but he needed to keep his own pills with him in his room. Peter woke up sleepily and took the rest of his medications. Dorothy came back from her scan, and I gave her a good look and listen. I also had time to do some charting on the computer. The rule is, if it isn't charted, it isn't done, so nurses are motivated to chart to protect themselves legally. Certain kinds of charting are required, but beyond that the amount of charting any one nurse could do on any shift is almost infinite, so charting tends to create a constant nagging feeling of "things left undone." When it felt done enough for the moment, I checked on the computer to see if the doctors had put in any new orders. It was eleven o'clock, and I was still pretty much caught up.

Then Dr. Arable, the infectious disease specialist, came to see Tom, and my day went briefly haywire. Many people say that Dr. Arable looks like an undertaker, and I might agree with this if I thought undertakers could afford the expensive suits and exquisitely patterned ties he wears. When I first started at the hospital, I found Dr. Arable reserved to the point of forbidding. A patient's husband even described him as "Sphinxlike." Over time, though, I saw his more human side, his commitment to the patients, his wicked sense of humor. I was standing at my med cart looking up things on the computer when he came out of Tom's room. "How's he doing?" I asked. Dr. Arable made a wry grimace. "Oh, he's just taching away in there," he said, meaning that Tom's heart rate was stuck in the 120s. "I think he might be unit bound."

I swallowed. I'm still learning how to do these negotiations, but here was the opportunity Karen had suggested I look for. "Do you think it would be a good idea to do that sooner rather than later?" I asked him.

Dr. Arable considered the question and then said, "I don't think that would be such a bad idea."

"Are you willing to go on the record with that?" I asked.

He thought about it again. "Yeah," he said, "let me wander across the hall and see if I can find the team and talk it over with them." He paused. "I just don't want to walk into his room and find him dead."

"What?!?" I wanted to scream at him, but instead I said, "Um, yeah. That would be a really big bummer." *Bummer* really was not the right word here, but I did not want to walk into Tom's room and find him dead either, and until that moment I had never considered it as a possibility. What I really wanted to say to Dr. Arable was, "What the hell are you talking about?" and I'm not sure myself why that came out as *bummer*.

About five minutes later Dr. Arable came back and told me he could not find the team, but that he did think Tom should go to the ICU. Now it was my turn to swing into action. The appropriate thing to do was to call the intern and let him move up the ranks as needed to effect an ICU transfer, but I had enough experience by this point to know that talking to the intern would most likely be fruitless, so I paged the fellow, Sujata. She said the transfer was fine with her if Dr. Arable recommended it, but she was in clinic and couldn't come over to the hospital to physically see the patient. I had no option now but to page the intern.

I had never transferred a patient to the unit and didn't know how it worked, but I knew it wasn't as simple as asking for a room. I explained the situation to the charge nurse and told her Dr. Arable said he didn't want to walk into the room and find Tom dead. "Well, yeah," she said, "because that happened to him."

Again I thought, "What?!?" and "Why doesn't anyone tell me these stories?" But instead I kept my cool and asked, "Was that the guy who died right before he was discharged?"

"No, that was someone else," she explained. "One time Dr. Arable went into the room, came out, and said, 'That guy's dead.'"

It took multiple pages of the intern and even Gayatri going

to look for him to get him to call me back. Then he wanted to order the transfer without physically seeing the patient. Finally, he came, saw the patient, and, somewhat coached by me, spoke with the intensivist, who was, again, Dr. Sutherland. Dr. Sutherland aggressively guards access to the unit. He kept the intern on the phone for almost ten minutes, an eternity in the hospital, but the intern must have done his job, because Dr. Sutherland agreed to the transfer. It could also have been because of Dr. Arable's recommendation. Until this moment I had never realized just how much pull he had on the floor and in the units.

The nursing supervisor called with a bed, and then everything had to happen at once. I had gotten behind on Tom's drugs while I was trying to arrange his transfer, and now I needed to get him caught up and move him and his many bags down to the ICU. Two more senior nurses stepped in. "Theresa, do you need help?" they asked. I think I had a sort of deer-in-the-headlights look on my face, and I mumbled, "Uh-huh," followed by lots of effusive thank-you's. I gave Tom his noon pills, his noon antibiotic, and his IV steroids, while the other two nurses hooked him up to an oxygen tank and piled his belongings on top of his bed and in a wheelchair. The lab technician still had not come to draw the tobramycin level even though it was now 12:30, so I called her back and left a message saying the patient was going to be transferred to the ICU and could she draw the level there.

Marlo wanted to take a shower, so she needed to be disconnected from her IV fluids. Dorothy's back pain was flaring up, and she wanted an oxycodone. She also needed her noon insulin and was feeling nauseated. I asked two other nurses to help. One knew Marlo well and was happy to disconnect her from her IV because it meant a chance for a chat. Another gave Dorothy the drugs she needed. Peter was fine for the moment, caught up and mostly just waiting for his ride while he played rock videos at high volume on his laptop.

Together the three of us got Tom rolling. In the maybe five minutes it took us to get him out of his room and onto the elevator, my portable phone rang three times. The first phone call was the lab technician calling me back to say she had been at lunch and could the ICU nurse draw the tobra levels. I wanted to say, "Lunch? You get a lunch?!?!?" but instead I just told her OK and thanks for calling me back. The second phone call was from the pharmacy telling me Dorothy's chemo would be slightly delayed. OK to that as well—I just had to remember to pass the information on to evening shift. The third call came as we were guiding Tom's bed, with the attached oxygen tank and his IV pole, onto the elevator. There's a bump when you get onto the elevator, and it can be tricky to get the bed across. This call was from Jack, the ICU nurse, who was saying he couldn't find my taped report. "We're just getting into the elevator," I told him, "but I did tape."

We were navigating the same bump getting Tom off the elevator when my phone rang again. It was Jack telling me that he still couldn't find my report. "OK," I said. "We're just getting off the elevator, and I will be there in less than five minutes and can give you a verbal." It is not so easy to push the back end of a hospital bed and an IV pump while holding a phone up to one's ear and trying to sound coherent. Karen, the nurse with the expressive eyebrows, laughed wryly each time my phone rang and said, "It's you again." Finally, we got there, and Jack again told me that he couldn't hear the report I taped. "You must have put in the wrong room number," he said.

This is the kind of comment that, if I were a cartoon character, would make my face turn red and steam come out of my ears. I bit back a "No, I did not put in the wrong room number," and instead looked directly at him. "What would you like me to do?" I asked him. "Oh, I can look up the history," he said in a much friendlier way. Now I was confused. Was he just giving me a hard time because he could, or was he genuinely worried that information central to taking care of the patient might have gotten lost?

Tom had not wanted to go to the ICU. Before I left him there, I once again told him, "Short stay probably, Tom. They'll get you a little more stable here, and then you can come back to us." I took hold of his hand. "That's what we'll wish for, Tom—a short stay."

Then I explained to Jack about the tobramycin level. I explained it three times in detail. If he remembered one thing I said, I wanted it to be the tobra level. I also told him, on my own authority, to hold the tobra until he had drawn the level.

Back on the floor I first checked that all my ducks were in order: yes, my three remaining patients were fine. Next, I called Gayatri, the pharm D, and told her that Tom had indeed gone to the ICU, that I had explained about the tobra level to the ICU nurse, and that I had held the dose of tobra because it seemed like getting the right level was more important than getting the dose exactly on time. "Oh, yes," she said, "it's OK to hold it—we really need the level." I gave another big internal sigh, this time one of relief: as a nurse, it's not really my call to hold an antibiotic, but it had seemed like the right thing to do.

Around this time I ate something for lunch while joking that one way to reduce your patient load was to send patients to the unit. My phone didn't ring once while I was eating, and I was glad for the serendipitous break. However, right as I finished my fifteen-minute meal, the secretary found me and chided me for not having my phone on. I checked it. The battery had died. The peace I got during lunch came at the cost of three missed phone calls. It had been a nice respite, but it had been illusory after all.

It was now 2:00 P.M., and I had ninety minutes until my eight-hour shift was done. Peter needed medicines for his discharge, and they were being sent over from two different pharmacies due to the requirements of his insurance. I also still needed to complete, print out, double-check, and sign his discharge instructions. Dorothy was taking what looked like a long, lovely nap. It is hard to get quality rest in the hospital, so I let her sleep.

Marlo now needed the bulk of my attention. She grew increasingly anxious as the time for the bone marrow biopsy approached. Pretty much everyone who has a bone marrow biopsy dreads the procedure. It is done at the bedside under a sterile drape. The fellow or nurse practitioner extracts a sample of tissue from the patient's hip bone by inserting a long needle into the bone and pulling out a "core" of tissue. The skin is numbed before the needle goes in, but the bone itself cannot be anesthetized. People seem to find it uncomfortable more than painful, but the discomfort is eerie, haunting. Bone, after all, is living tissue, and patients feel the needle go in, the turning needed to extract the sample, and the pulling to get the needle out. Sometimes it takes more than one try to get a good sample, and patients find that uncertainty difficult as well.

Sonya, the fellow who would be performing the biopsy, asked me about anxiety medication. "She wants IV Ativan," I said. "OK," she said, "give her 1 milligram."

I drew up the drug and went to give it. Sonya was in the room talking over the procedure with Marlo, and Marlo kept talking about how anxious she felt. Then she told us, in her quiet voice, "You know, it was when Dr. . . ." (I couldn't catch the name) "did it, he didn't give me anything, and that was terrible."

"What?" I looked at her. "No, that's ridiculous," I said. "You just need to get hooked up with the right people, Marlo. We'll take care of you." She, Sonya, and I all laughed together, but it was really laughter from relief. She was relieved to have told her story, and I was relieved to understand why her anxiety was so extreme: having a bone marrow biopsy done without pain or anxiety medication given beforehand to take the edge off would leave a patient very fearful for the next time.

I gave her 1 milligram of IV Ativan, and then another 0.5 milligrams thirty minutes later. We have to take special vital signs for IV Ativan, and the vitals have to be taken at precise time intervals. Since I gave two doses at two different times, I ended up taking a lot of extra vitals. Marlo had three people in the room with her, including her sister, so I knew she wouldn't feel alone, and I left to discharge Peter.

I got Peter his discharge papers and his meds, but before we did all that, he insisted on playing me some Springsteen. I already knew we had surprisingly similar tastes in movies, and I had confessed to him that I did not like The Boss. He played me some early Springsteen and then a song from his album *Magic*. I had told Peter that I didn't like *Born in the USA,* and he snorted as if to say, "You think *that's* Springsteen?" This new song I liked, and I stayed in his room the entire seven minutes listening to both songs. I tried not to look at my watch, I put aside thoughts of what Marlo might need, and I didn't think about the charting I still needed to do. When the music finished, I realized Peter had given me a tremendous gift; he had allowed me to be a regular human being. Not his nurse—his pill giver, paper distributor, and vitals taker—but a person, maybe like the mother he had already lost to cancer, maybe even a little bit like a friend. When the songs finished, the spell was broken, but the magic lingered for me. I hugged him and said what we always say to patients when they leave: "I hope I don't see you for a long, long time."

Marlo seemed miserable whenever I checked on her during the procedure, but when it was over, and I asked her how she was, she smiled tentatively and said, "I think I'm OK. I really think I'm OK." I got Dorothy caught up and then went to pick up papers for the transfer I was getting from the ICU. This patient was going into Tom's old room, and I joked with my boss about playing musical chairs with the unit.

Marlo's biopsy site bled onto her sweatpants, but not an amount of blood to worry about. Still, she needed to get herself cleaned up and her dressing changed. Sonya asked me if I thought some Ativan by mouth was a good idea for the next few days since Marlo seemed so anxious. I told her that was a good idea, since the biopsy result would tell her if we had gotten rid of the cancer this time around.

The end of my shift, 3:30, came and went, and I finished by typing Marlo's extra vitals into the computer. "Theresa, you're still here?" people asked. "Yeah, yeah, I'm leaving." I realized I hadn't told Jack, the ICU nurse, about Tom's Coreg and Zocor, so I called him. He once again mentioned the missing report, but at the end of our conversation he also said, "Thanks for all your help." When I was finally done, I threw my papers, which were now useless and illegal, into the paper shredder in our break room.

I changed into regular shoes, got my bag out of my locker, and swiped out.

Dr. Arable stepped onto the elevator just as I did. "Good job getting that tobra level done," he said. "It doesn't get done right very often." And suddenly the whole day seemed worth it—I had done something well after all; I had even been effective.

Because here's the truth—these four patients are all now dead. Tom went first, dying from acute respiratory failure just a couple of weeks after this day, and after several more trips back and forth from the ICU. Peter died rather suddenly a few months later. He went septic, and the shock killed him. His death hit me hard. He had already lost his mother and father to cancer. When I was his age, my biggest worries were whether I would get an A on the paper I had just turned in and how I was going to spend my weekend. Now he's dead, and the most I can do is listen to Springsteen and remember him when I do. And Dorothy, who had been doing so well, is dead now, too. She was sailing along when one complication after another hit her like tidal waves. By the time she died, she had so many things wrong with her that I'm not sure what actually killed her. And Marlo had died between the writing of this book and its publication. In the end, her chemo had irreparably damaged her heart.

So, a sincere compliment, a brief sampling of good Springsteen, my first successful transfer to the ICU, some jokes about providing Marlo with IV benzos, and at the end of my shift they were all still alive. During my eight hours that day I tried to help them in whatever way I could. For our patients, there may not be a tomorrow—today has to count. At the end of the shift, I went home, saw my kids, ate dinner, slept, and showed up at 7:00 A.M. the next day, ready to do it all over again, this time for twelve, not eight, hours. Another day on the floor.

SIX

Condition A

Usually when patients die it's at the end of a long struggle—we've done everything modern medicine can do and then some, but we can't save them. Some part of their body, their lungs or heart, possibly their liver, has become too frail, too vulnerable to function without the support of tubes and machines and drips and drugs that can keep them living for a while, but will not save them. These are the "good deaths," the ones where the family is present and knows, as much as possible, what to expect. Like all deaths, these deaths are difficult, but they are controlled, unsurprising, anticipated.

And then there are the other deaths: quick and rare, where life leaves a body in minutes and never comes back. In my hospital, these deaths are referred to as "Condition A's." The *A* stands for *arrest,* as in cardiac arrest, as in this patient's heart has all of a sudden stopped beating, and we need to try to restart it. As in this patient is dead.

It's an unspoken rule that staff do not run in hospitals, but for a Condition A, people run, or at least they move very fast. Just a few weeks after finally finishing orientation, after coming back to the floor following my knee injury, I had my first Condition A. Then, a couple of months later, I had my second. Some nurses go their entire careers without ever having a Condition A. Having two in three months reflects bad luck, but also shows that nursing is a crapshoot: at the start of a shift, you have no idea what the work day may bring.

For the first Condition A, my patient was a middle-aged woman with lung cancer. She was very personable, vivacious even, with a wide, friendly smile and thick, long hair that stood out among our many bald patients with cancer. She was, as we say, "stable," "fine," still experiencing the same low-grade fever she'd had off and on for a couple of days, but well enough that her discharge was planned for that afternoon. She had come in because she was coughing up blood, a problem we'd resolved. During my morning assessment of her, we talked about whether she needed pain medication, and I promised to bring the Tylenol I had asked the intern to order for her fever.

I left her in the care of the nursing student who had her for the day and moved on to one of my other patients, thinking I was going to have a relatively calm day. About a half hour later, the aide called me: "Theresa, they need you in 1405." I stopped what I was doing and walked over to her room. The nurse leaving the room said, "She's spitting up blood," and waved to me from the nurses' station. "I'm calling her doctor." Inside the room I saw that my patient was indeed spitting up blood—not vomiting up blood, but spilling it out of her mouth and nose with absolutely no control. I remembered to put on gloves, and the aide handed me a face shield. I moved closer; I put my hand on her shoulder. "Are you in any pain?" I asked, thinking that a GI (gastrointestinal) bleed would be more fixable than whatever this was. She shook her head no.

I looked in her eyes and saw . . . what? Panic? Fear? The abandonment of hope? Or sheer desperation? Her own blood was gurgling in her throat, and I yelled to the student, "Tell them we need a Yankauer. We need suction." The patient tried to stand up, and I told her, "No, don't stand up." Blood was pulsing out of her mouth, and she was trying to get the bulk of it into the nearby trash can. She sat back down on the bed and then started shaking. "Is it condition time?" asked the other nurse in the room with me when the patient collapsed backward on the bed. "Call the code!" I yelled out. "Call the code!"

The next few moments I can only describe as surreal. I felt for a pulse, and there wasn't one. I started doing CPR. I heard the code overhead: "Condition A." The other nurses from my floor came in with the crash cart, and I got the board. Doing CPR on a soft surface, like a bed, doesn't accomplish much; you need a hard surface to really compress the patient's chest, so every crash cart has a 2-by-3-foot piece of hard fiberboard for just this purpose. I told one of the doctors who had come to the code to help pick her up so that I could put the board under her: she was now dead weight and heavy. I kept doing CPR until the condition team arrived, which seemed to happen faster than I could have imagined: two intensivists, two ICU nurses, a respiratory therapist, and I'm not sure who else, maybe a pulmonologist, maybe a doc from anesthesia. Respiratory took over the CPR, and I stood back against the wall of the room, bloody and disbelieving. My coworkers did all the grunt work for the condition, put extra channels on the patient's IV pump, recorded what was happening, and every so often called out "Patient is in asystole again," which means that she had no heartbeat.

They worked on her for a half hour. They tried to put a tube down her throat to get her some oxygen, but there was so much blood they couldn't see. Then they eventually "trached" her—

put a breathing hole in her neck right into her trachea, her airway—but that filled up with blood as well. They gave her fluids and squeezed bags of epinephrine into her veins to try to get her heart to start moving. The rap on a true cardiac arrest is that drugs cannot help because there is no cardiac rhythm for the medications to stimulate, but the doctors tried anyway. They went through so many drugs that the crash cart got emptied out, and runners came and went from pharmacy bringing extras.

I doubt very much that anyone in the room thought we could save this patient, could bring her back to life. We hoped, or at least I did, even though at some level I knew that hope was futile. The truth is, we don't like deaths like this in the hospital: sudden deaths, gruesome deaths. We want to feel in control, to know, or to think we know, that someone's body will have the usual and expected reactions. There is a word for what happened to my patient—exsanguination—but the word itself does not capture the experience of looking into someone's eyes as her life literally bleeds out of her.

When actors go through these routines on TV, it seems exciting and glamorous, but in real life the experience is profoundly sad. "Asystole" is known in the lay vernacular of Hollywood movies as "flatlining." But my patient never had the easy narrative of the normal heartbeat that suddenly turns straight and horizontal. Her heartbeat line was wobbly and unformed, occasionally spiked in a brief run of unsynchronized beats, and at times looked regular, because chest compressions from CPR can create what looks like a real cardiac rhythm on an EKG (electrocardiogram) even though the patient is dead.

At one point during the condition, seeing what looked like a normal heart rhythm on the monitor, I turned to Maria, the short, often bubbly nurse standing next to me, and whispered, "Is that a heartbeat?" Watching that heart rhythm—normal sinus rhythm—on the monitor, I felt briefly, but profoundly, exhilarated. This feeling, I understood, was a disbelieving bliss, the only possible reaction to witnessing a true miracle. Maria looked at the respiratory therapist doing chest compressions, looked at the monitor, and then looked at me. "CPR," she said. "Damn," I said, and then, "Damn, damn, damn, damn, damn."

Because my patient was dead. She had been dead when she fell back on the bed, and she stayed dead through all the effort to save her, while blood and tissue bubbled out of her, and the suction clogged with particulate matter spilling from her lungs. We did what we knew how to do to save her because that's how we're trained. She could not be saved. The reigning theory was that a portion of her tumor had broken off and either ruptured her pulmonary artery or created a huge blockage in her heart. Apparently this can happen without warning in lung cancer patients. Only an autopsy would tell for sure, and in terms of the role I played in all this, it doesn't matter. I did the only thing I could do—all of us did—because it was the only thing we knew to do. If a patient is exsanguinating, or bleeding out, with no pulse, and CPR is the only trick you have up your sleeve, plus or minus a code team and a cart full of drugs, the chances of a recovery, or to label it honestly, a resurrection, are slim to none.

Why do it then? Why even call a code? The truest answer is that legally we had no choice. Unless a patient has made it clear that he does not want emergency resuscitation, nurses and doctors are legally and ethically required to do all they can to save that patient. "Made it clear" must also be done officially. In my hospital, patients indicate their "code status" on bright pink sheets of paper that get placed in the front of their paper chart. These forms are signed not by the patient, or by the nurse who is taking care of the patient that day, but by a doctor. The doctor may be a very experienced attending, but might also be a twenty-three-year-old intern with very little sense of how to talk to patients about their own deaths. When it comes down to authorizing the doctor to sign the code status form, the patient herself may balk at signing off on any chance of being saved. The doctor may be too rushed to sit down with the patient and discuss her poor prognosis honestly. The form itself, and what it implies, may be deliberately ignored since some physicians believe that bringing up "code status" contradicts the idea that patients come to the hospital to be treated, get better, and go home. However, regardless of the reason, if a patient's chart lacks a signed code status form, the nurse must call the code.

Calling the code was the easy part; living with my patient's death was much harder. I had blood all over my white scrubs and no other clothes to change into. One of the ICU nurses who had been at the code brought me a set of teal scrubs, the kind worn by respiratory therapists, and said, "Don't ask where I got these, just put them on." In our staff bathroom I took off my bloody pants, scrub top, and white cotton shirt. The pants and shirt I knew I would throw away, but the scrub top had just a small smudge of blood on it near the bottom hem. I really liked that scrub top. It was a mix of blue and teal dots on a white background, and people always said it looked happy (even though it violated the all-white dress code). I washed out the bloodstain from my top while I stood there in front of the small sink feeling bereft. My patient had been so alive just minutes before, and now the en-tirety of her life had condensed for me into the stain I had to scrub out of my shirt. I stuffed my bloody pants and bloodstained white long-sleeved shirt into a red biohazard bag: the only memorial I could offer her.

Somehow I got through the rest of my twelve-hour shift and did another twelve-hour shift the next day as well, but I felt jumpy at work and depressed at home. My boss's boss, Sarah, a completely no-nonsense woman trained as a nurse anesthetist, had sought me out later that same day and called me into my boss's office for a private talk about what had happened. She told me that nurses who have experiences like mine are able to keep working if they find a way to channel those experiences into the good things in their lives: family, kids, home. I believed her, but days went by and the feeling of having a pall over me did not lift.

Solace came finally from an unexpected place. I had made a casual plan to see the movie *The Counterfeiters* with my friend Judith's husband, Daniel. The German title is *Die Fälscher*, and the movie tells the true story of Jews who were skillful forgers interned at Sachsenhausen concentration camp outside Berlin as part of a secret program to counterfeit the British pound and the American dollar during World War II. The movie begins, and we see Salomon Sorowitsch, the pragmatic criminal and counterfeiter who becomes the movie's moral center, trapped and caught by the Nazis. Marching into his first concentration camp—before he was transferred to Sachsenhausen—Salomon watches as a Kapo, one of the prisoners-turned-guard, beats another prisoner to death.

Watching this scene, all I wanted to do was turn away. Not more violence, I thought, for my patient's death had been violent, and certainly not more blood. Then it came to me, not the idea that "things could be worse"—because I agree that being in a concentration camp would be "worse" than having your patient bleed out in Pittsburgh—but that the relative worseness of particular situations is irrelevant. What comforted me while one prisoner beat another to death in *The Counterfeiters* was the realization that with so much cruelty in the world, all any of us can do is try to make our own piece of it better. The violence and the cruelty, the illnesses and the blood, will not stop, but I can put my hand on my patient's shoulder and ask if she's in pain. I can look in her eyes as she dies and then try to pump life back into her. I can offer up my own hope, regardless of how futile that hope is, and whenever it's needed, keep offering it. I can, within the limits of my own life and limb, always try.

The movie continued and I fully grasped the importance of trying, always, to care, when my own memories of Sachsenhausen intruded on my cinematic experience. In my midtwenties I visited Sachsenhausen concentration camp the summer after the Berlin wall fell. The camp had been turned into a museum of sorts. It was located on the far northern outskirts of what had been East Berlin and was the last stop on one of Berlin's suburban trains.

The walls around the camp still stood, the smokestacks of the crematories were there, and a brochure or guide, available in many languages, could be purchased for a nominal fee. The brochure outlined a directed tour that included the chimneys and the basement of the "hospital" where inhuman "experiments" were conducted on human beings. The guide had been produced by East Germany when there still was an East Germany and made much of how greatly the communists in the camp suffered, especially when contrasted with the comparatively milder suffering of the Jews.

The camp "shoe track" was also featured on the tour, and the brochure explained that guards in Sachsenhausen would force prisoners to put on shoes too small for their feet and then make them run around and around the "shoe track." At the time I visited Sachsenhausen, I did not believe the story of the shoe track. Who would do such a thing? I wondered. And why? I dismissed it as bizarre East German propaganda comparable to the claim that communists in the camp were treated much worse than the Jews. But in *The Counterfeiters* when Sally, as the other prisoners call him, begins his work with the forgers in Sachsenhausen, he questions why they shut the windows against the noise from the shoe track. He's told that the shoe track is used to torture prisoners, and that prisoners even die while running on it. So, almost twenty years after my visit to Sachsenhausen, and just a few days after my patient's death, my own disbelief in the existence of pointless agony, of needless suffering, was shattered. The blood on my clothes in the hospital, the memory of the look on my patient's face, and the truth of the shoe track came together in that moment, confronting and changing me.

My second Condition A was completely unlike my first. For one thing, I wasn't even there. My patient, Robert, died in the hallway following a radiation treatment. The technicians had put him in the hallway to wait for transport to take him back to the floor, checked on him five minutes later, and discovered that he was dead, or as we say, that he had "arrested." I was on evening shift that day, doing a three to eleven, and my shift had just started when we heard "Condition A, Radiation Oncology" over the loudspeaker. I had not seen Robert yet that day because he left for his treatment before my shift started, but I knew him from previous admissions. When the Condition A was announced, several of us who were standing at the nurses' station stopped and took a few seconds to think, are any of my patients in radiation oncology? I knew it could be Robert who had died, but despite the seriousness of his condition, it seemed unlikely. It seemed unlikely in part because he did not have a history of heart trouble, but to me it seemed most unlikely because I just had a Condition A two months before. As I saw it, the odds were in my and Robert's favor. However, sudden death has no regard for probability, and a few minutes later our unit secretary told me in whispers that it was Robert in the Condition A.

"Oh my God," I said, over and over and over again. I knew this man, and I knew his wife. They were both in their late sixties with adult children and grandchildren. He was incredibly nice, and her concern for him took the form of fussing hyper-vigilance. She could be intense and demanding, so much so that some of the nurses on the floor found her annoying, but it was obvious that her worry was motivated by love. She demanded food that Robert would actually try to eat, she tried to arrange his schedule of tests and treatments so that he could get some rest each day, and when he became confused for a while and didn't know where he was or who he was, she made sure we had enough staff to get him safely to and from the portable commode at his bedside.

She was always with him, always at his side . . . except for this moment on this day, when he died in the hallway. I found myself wondering what she had hoped to do during her short break while Robert went to radiation. Was she trying to enjoy a quiet cup of tea in the hospital cafeteria? Did she need to call and update the children on how their dad was doing? I hope she was not on the phone arguing with their insurance company about copayments and coverage, but she may very well have been. Regardless, no one knew where she was, so we paged her: "Mrs. Robert List, please return to your room on the fourteenth floor." Our charge nurse, Anna, a heavyset blonde with an at times explosive personality, was called down to radiation oncology, where the code team was running the "condition" on Robert. My job was to wait on the floor for Mrs. List, give her the news about her husband, and escort her down to the code myself.

In nursing school I received absolutely no training on how to tell a woman her husband of thirty-plus years has died in the hallway during the very brief period she had left him alone in the hospital. I received no training on talking about death to anyone, and this situation taxed me. What does one say? I started with introductions. I knew that she knew me, but it seemed important for me to know her as well. "We've talked so many times," I said to her gently, "but I've never learned your name. What is your name?" "Marianne," she told me, her face looking both apprehensive and blank, as if her worry was so great that some protective part of her brain had already forestalled it.

It was early evening by this time, and the room was dark. She had not turned on a light when she returned to the room, and I, equally as distracted, had left the lights off when I followed her in. It was dusk outside, and the darkness of the room contrasted with the well-lit hallway just beyond the door. "They need you in radiation," I told her. "What?" She was a thin woman, casually but expensively dressed, and without makeup. Her face was framed by soft brown hair that contrasted with the angular presentation of her cheekbones, the worry that was pulling her skin tight. "What happened?" she asked me, and her voice did not break, but it echoed the tightness of her skin, the lines on her face pulled taut like strings on a violin.

I paused mentally even though I had no time to consider my words. I was not actually thinking about what to say as much as I was struggling with the truth. I fought an incredible temptation to lie. My mind did not want to release the words "dead," "nothing to be done," "not breathing" and instead wanted to offer up comforting misrepresentations, "We don't know what happened," "The code team is with him now," "We'll see when we get there." Looking at Marianne, I seemed to be viewing her through water and wondered how I could penetrate such a dense medium. And then I won my battle with the urge to lie, and the space between us cleared, and I found the best words I could: "They called a Condition A on your husband, which means that he had a cardiac arrest."

She looked right at me. "What does that mean?"

"His heart stopped beating. They found him in the hallway." She looked at me, unsure of my meaning because, I now saw, she could not understand the content of my words. Husbands do not die while one makes a phone call, has a quiet cup of tea.

"They need us in radiation," I told her again, and began walking her out of the room and toward the elevator. My phone rang; it was Anna. "Theresa, where are you?" "We're walking to the elevator," I told her. Once we were on the elevator, going down to radiation on the first floor of the hospital, my phone rang again. "Theresa, they really need her down here." "Yes," I said, "we're on the elevator," trying to sound calm, but wanting to shout in frustration: "I can't speed up the elevator! I can't get us down there any faster." Marianne looked at me, clear-eyed. "Is he gone?" I felt the water come down between us again as half-truths and pieces of lies swirled around in my head. I so wanted to tell her the truth—the least we could give her was the truth. I looked at her. "He could be," I said, because we wouldn't know for sure until we stepped off the elevator and saw Robert for ourselves.

In my hospital you have to walk down a large open hallway to get from the elevator to radiation, and Anna and my boss's boss, Sarah, met us there, near an information desk, where transport staff, doctors, nurses, and worried relatives of other patients passed by. Sarah, who counseled me after my first Condition A, is very direct; she had one question for Marianne and wanted her to answer it. Marianne, however, had her own question, and she wanted Sarah, or someone, to give *her* an answer. She wanted to know if Robert was dead, and Sarah wanted to know if they should stop doing CPR. Each of their questions was important, but Marianne would not answer Sarah's question without getting an answer to her own, and Sarah felt she could not answer Marianne's question without getting an answer to hers. Robert was dead and had been when they called the code—that's what a Condition A means—but neither Sarah nor Anna nor I could say, "Yes, he's dead," until they stopped the code. It was a grotesque catch-22. Marianne could only be told that Robert was dead once she had authorized us to say he was dead: to "call" the code. It was Marianne's decision to make legally, but forcing an answer from her in this situation seemed cruel.

She herself must have sensed that, for she refused to answer Sarah's question and would not be deterred from her own. "Is he gone?" she kept asking, standing in the hallway while people walked by, oblivious to her worries, focused on their own. The look in her eyes grew increasingly frantic as Sarah in turn refused to answer Marianne's question. I felt the same frustration I had felt in the elevator: "Just answer her question!" I wanted to shout. I also felt trapped, because nursing has an often unacknowledged but rigid hierarchy, and flanked by my charge nurse and by Sarah, I did not feel I could intervene in what they considered a necessary, if galling, procedural formality.

Indifferent to our procedures and formalities, Marianne refused to answer Sarah, and Sarah finally grasped this. She gestured away from the hall, toward radiation, where the code was still taking place, by which I mean they were still doing CPR, still giving him oxygen, maybe even administering drugs. Marianne followed Sarah's gaze, walked down the hallway, through the open doors to the radiation department, and through the closed double doors that kept the code private. She caught a quick view of her husband concealed beneath a cluster of people all very focused on doing their jobs. Michael, one of the most gentle intensivists was "running" the code: giving directions to the staff—nurses, respiratory, anesthesiology—who worked in concert to give life support to someone who could not be saved. Oh my God, I wondered; had they been doing this for forty minutes?

Marianne turned her back on the code and burst back out the closed double doors and into the entryway to radiation. "I can't stand it," she said. "I can't watch him suffer." This seemed to be enough of an answer for Sarah. She stepped back through the double doors herself, and Anna and I walked Marianne into the waiting room that was just inside the entrance to radiation. It was an odd little room. If you turned left, a narrow anteroom was taken up with four chairs set in a row against the near wall, and the opposite wall opened up into a slightly larger area that had lockers for clothing along one wall and, improbably, a sink along the other.

Marianne could not sit down in the waiting area. She paced in the small space between the lockers and the anteroom, clutching her cell phone. She had already called her son who lived nearby and asked him to come to the hospital. Now while she paced, she spoke with her daughter who lived out west. "Daddy's gone," she kept saying, but it was clear that her daughter had the same trouble understanding what had happened that Marianne herself had initially. In frustration she handed me the phone—"Here, Theresa, you explain it to her"—and sat down in one of the ante-room chairs. I took the phone and heard an unintelligible wailing from her daughter. The daughter did not want to talk to me and in fact could not talk to me. Her wailing, I realized, was the sound of raw grief, a sound so true and so painful that hearing it feels like a smack across the face. I felt lost in that sound, and helpless. "You want to talk to your mom, right?" I finally managed to ask, feeling stuck between wanting to help Marianne and knowing that my presence on the phone was tormenting her daughter. The daughter blurted out, "Yes." I handed the cell phone to Marianne, but it turned out that something had been accomplished by my brief turn on the line. Marianne's inability to talk to her daughter had made the daughter understand that her father was dead.

Anna, Sarah, and I all stood in that anteroom listening as Marianne talked to her daughter. Time seemed both to stop and to extend into eternity: as if we had always been in that room and would never again be anywhere else. My phone rang, and it was a shock to hear the voice of our unit secretary, "Theresa, are you still down there?" I didn't know how to answer her; where else would I be? The hurly-burly of the hospital had no traction in that odd waiting room. "Can you tell Anna that we're getting killed with admissions?" For the third time that evening I felt as if I were talking through water. The secretary's voice and concerns seemed so far away, so trivial, contrasted with Marianne's nonstop crying. Nevertheless I told Anna, who waved away the secretary's news with a flip of her hand, saying, "Not now."

Robert's primary oncologist, Dr. Souchet, came into the room next, and Marianne stood up and put her arms around him. "Oh, Dr. Souchet," she cried, dwarfing him since Dr. Souchet is quite short. The two of them sat down together, while we three nurses continued to stand and look on, a Greek chorus for Marianne's tragedy. He did not say to her, "It's better this way," because having a husband die is never "better" than the alternative, but he did say to her, very quietly, that they weren't sure if the next scheduled treatment would have done more harm than good and that Robert had been weakened by the treatments—the chemo and radiation—he had already had. She nodded; she understood that his death had been coming, just not this evening, just not today.

My phone rang again. "Oncology, Theresa," I said by instinct. It was an ICU nurse from the code wondering if the family wanted an autopsy. The question mattered because the nurses were cleaning up "the body," as Robert had now become, and having an autopsy would mean that all the tubes that had been inserted during the code would have to stay in until he got to the medical examiner's office. Without an autopsy the IVs and tubes could come out, which would make seeing Robert's corpse much easier on the family. Do we have to decide this *right now?* I thought. I didn't know what to say: Marianne and Dr. Souchet were having their quiet talk, and I couldn't imagine intruding with this question. I inclined my head to Sarah and told her in an undertone, "They want to know if the family wants an autopsy." I handed Sarah my phone, and she talked to the nurse from the ICU, but the autopsy decision could not be postponed, so we asked her.

"What?" That frantic look came back to Marianne's face. "An autopsy? No . . . I don't want him cut open." But she wasn't sure; she had said no, but doubtfully. We wanted her to be sure, either way, and we all began talking at once: "It's up to you," "Your decision," "No right or wrong answer." I don't think she cared one way or another about the autopsy, but the decision was once again hers alone to make. I spoke up amid the cacophony of voices, picking my words with care. "Marianne," I said, "some people feel better knowing what the exact cause of death was—it comforts them. Other people don't need that knowledge; they can find peace without it." She looked at me with an earnest yet vacant look, as if she wanted to understand what I was saying but couldn't. Then she spoke, waving her hands in front of her as if she were chasing away gnats. "No, no autopsy—I don't want him being cut up."

The whole thing went on and on. Dr. Souchet left, and I took Marianne back up to the floor. She had to wait until they brought up Robert's body. I made sure there were tissues in the room, and when the body was ready, and her son had arrived, I took them into the room and left them there to be alone with Robert. The death certificate had been completed by one of the doctors at the condition, and just one task remained for me—to contact the funeral home. Usually a nurse makes this initial call to the funeral home and follows their direction as to whether the family needs to call also and how soon they will pick up the body. The rule is that cadavers are not supposed to stay on the floor for more than two hours, and if the funeral home is far away, or if the family cannot bear to leave the hospital, meeting this time limit can be tricky. Somehow when I call I always expect a sad, sighing voice on the other end of the line, but invariably the directors are crisp, professional. To them death means business, unfortunate but very routine.

Since I came to nursing circuitously following my brief career as an English professor, I often used to hear John Donne in my head at the hospital: "Death, be not proud, though some have called thee / Mighty and dreadful, for thou art not so." Donne concludes his famous sonnet with "Death, thou shalt die," calling on his Christian belief in an afterlife to strip death of its power. The imagery is compelling, but after my two Condition A's, I find his words meaningless. Death is always death, and in real life, especially in the world of the hospital, sudden death, whether violent and gruesome or unbelievably prosaic, is unsettling.

What can one do? Go home, love your children, try not to bicker, eat well, walk in the rain, feel the sun on your face, and laugh loud and often, as much as possible, and especially at yourself. Because the antidote to death is not poetry, or miracle treatments, or a roomful of people with technical expertise and good intentions—the antidote to death is life.

SEVEN

Openings

My patient that day was young enough that I thought of him as a kid, although nothing about his mien, or the physical ordeal he had been through, was at all childlike. He had a biblical name, Abraham, that went well with his history. He'd had a small abdominal tumor he spent a year trying to heal with prayer. During that year his tumor grew so large that the pain eventually drove him to the emergency room. The surgeons operated to remove it along with part of his bowel, which left him with a colostomy and a large abdominal opening. The tumor had been big enough that, paradoxically, once they took it out, he didn't have enough skin to suture the opening back up again.

I'm not sure why, but I liked him, even though most of the nurses on the floor thought he was a pain in the neck. He wanted his IV lines arranged just so, his pills given in a certain order, his table at a specific angle relative to the bed. He refused to have vital signs taken during the night, and some of our nurses found this an unacceptable health risk, or an unacceptable challenge to their authority, or both. They and he kept up a guerrilla war of insisting on and refusing vitals and arguing over where the IV pole and the table should be. When he got particular with me about aspects of his care, I just sassed him right back, and he took the teasing OK.

I felt sympathetic. Here he was, in his early twenties, with a colostomy, a gaping wound in his abdomen, and a struggle for his life that had been made much harder by the year he spent delaying treatment. He had tried to control his illness with prayer, and that choice had been a complete failure. Now his disease—and the inpatient care that disease required—controlled his life to a degree he found intolerable. So he wanted his table positioned a certain way, his pills doled out according to *his* specifications. There are patient quirks that drive me crazy, but Abraham's I seemed to understand intuitively, though I couldn't put my finger on what about him or me made that understanding possible.

His abdominal opening had been packed and dressed by the wound care team at the hospital, and I wasn't supposed to mess with it. For complicated surgical wounds, this special team sometimes does all the care—floor nurses on medical floors just don't have the expertise to use the available high-tech dressings and bandages the right way. His abdominal opening had been covered with a wound vac, which is about as high tech as wound care gets. The wound vac requires that special highly absorbent, dense sponges be placed in the wound. Then an airtight dressing gets put over the sponges, and part of that dressing connects to a small machine that sits by the bed, plugged into the wall, generating mechanical suction. The sponges draw up moisture from the wound itself, and the machine gently sucks that moisture out, the idea being to create a vacuum that will actually speed healing. Recovery rates for large wounds treated with wound vacs are close to miraculous—wounds that previously took four months to heal can close up in two, and wounds that simply would not heal, sometimes will heal with a wound vac.

Abraham ended up on our floor, a medical oncology floor,

because he was getting chemo. Because he still had a large wound and a colostomy, he wanted nurses experienced in caring for surgical patients to attend to him. It wasn't an unreasonable request, it just wasn't practical, and even though he knew that, being stuck on a medical floor really frustrated him. Due to our general inexperience with wound vacs and brand-new colostomies, Abraham thought most of us were fools. Maybe I didn't mind him as much as some of the other nurses because I myself felt pretty ignorant about wounds and dressings. I didn't mind going slowly and listening to his advice.

Despite my ignorance of most things surgical, that day I had to arrange a meeting between the wound care nurses, Abraham himself, and the plastic surgeon who had been consulted in the case. Abraham's regular surgeon wanted someone in plastics to check out his wound and decide whether plastics needed to be involved already or should wait until the wound was further along in the healing process. For me this meant paging and talking with the plastic surgeon to find out when he could be on the floor and coordinating his schedule with the nurses in wound care. The surgeon wanted the nurses to remove Abraham's dressing so that he, the surgeon, could examine how it was healing. It took some time to remove the wound vac dressings, and the nurses knew the surgeon would not want to wait in Abraham's room while they did that. After the surgeon examined the wound, the nurses would repack it with the special wound vac sponges and reattach the suction device to Abraham. The wound care nurses wanted to leave Abraham's wound open for as short a time as possible. This consideration made timing the unwrapping tricky and meant that the surgeon would have to be punctual.

At that time I found surgery and surgeons more than a little scary. The idea of curing people by cutting them open made intellectual sense to me, but emotionally I found the concept difficult to accept as "therapeutic." To take a knife and purposefully open someone's body flies in the face of what we think of as healing. The cut hurts, flesh bleeds, and the risks of an infection, an accidental nick, are always there. Skin is supposed to stay intact to protect our insides from the outside world. Slicing through skin, connective tissue, muscle, even to remove something as surely deadly as a cancerous growth, is dangerous, and it made me deeply uncomfortable.

To match the disquiet I felt about surgery, I also had a very stereotypical view of surgeons as arrogant, men-of-few-words types. My image of surgeons had nothing to do with surgeons I had actually met, but like all dumb stereotypes, mine persisted. It made me reluctant to page the surgeon. Even though he had the very generic name of Anderson and seemed really just like anyone else during our brief phone conversation at the start of my shift, I worried that he could pounce at any time and that my inexperience in nursing, as well as my ignorance of surgical procedures, would make me especially vulnerable.

By early afternoon the wound care nurses were in Abraham's room, and Dr. Anderson was supposed to be on his way. In another brief but completely ordinary phone conversation, he had told me he could be on my floor and in Abraham's room at 2:00 P.M. We all knew that Dr. Anderson wanted the wound open when he arrived, and the trick was not to remove the dressing so early that Abraham would have to lie there with his insides fully exposed for too long.

Dr. Anderson had assured me that he would be on time, so

the two nurses from wound care began to remove the dressing. Taking a dressing off a wound like this hurts, and I had given Abraham a dose of IV Dilaudid, a painkiller ten times as strong as morphine, to help make the pain manageable. Once they got the dressing off, I was able to peek briefly inside. I think I expected to look right into his abdomen and see his small intestine in a curling mass, but of course a layer of fascia, part of the body's connective tissue system, stretched across the entire wound opening, which extended vertically about seven inches from sternum to navel, and horizontally three inches side to side. The edges of the wound looked pink and healthy, but I only got a quick look because almost as soon as the nurses removed the dressing, they covered it again with sterile gauze. The nurses wanted to limit the wound's exposure to whatever pathogens might be in the air, exhaled as we breathed or circulated through the air vents or unwittingly kicked up from the floor.

After ten minutes of waiting, it became obvious to all of us that Dr. Anderson had been delayed. The more senior wound care nurse, with her calm face and long hair in tight curls pulled back into a ponytail, gently suggested that I page the surgeon one more time. I paged Dr. Anderson with some hesitation, but he called me right back and as soon as I said my name answered, "I'm on my way," and hung up. Wow, I thought—the surgeon was apologizing to me, and suddenly I felt really happy. I felt part of a team—an experience rare in university life—and a team in which we all had the same goals. Abraham wanted his wound to heal as quickly, neatly, and cleanly as possible, and the rest of us wanted that, too. Dr. Anderson's "I'm on my way" was not rude, just efficient. The sooner he got off the phone with me, the sooner he would be in Abraham's room. We're all just people here, I thought, and felt newly hopeful about one day doing the job with more confidence than I often had during that first year.

Possibly due to that burst of hopefulness, combined with relief at learning that Dr. Anderson was indeed on his way, we started talking in the room: Abraham, the two wound care nurses, and I. In nursing school I was taught to keep personal disclosures to a minimum with patients. This prohibition incorporates another stereotype, that of the saintly nurse with the kind smile, a very businesslike manner, and no needs or interests of her own. But Abraham's cell phone had rung in the middle of the wound unpacking. He didn't answer it, but this intrusion of his everyday life into our hospital work ended up triggering our conversation, a very ordinary discussion about cell phones and how they had really changed the social landscape. Abraham was in his early twenties, and the other two nurses and I were in our forties or older. I piped up with the observation that, "Yeah, my twenty-year-old friends from nursing school think it's so strange that I have a landline. They only use their cell phones." And everyone laughed and said they understood.

My comments were not profound, therapeutic, or even relevant, but I figured out something important after making this very small self-disclosure. I realized that patients with cancer do not want to be taken care of by saints. They are hospitalized for days, weeks, sometimes months. Those long lengths of stay pretty much guaranteed that the kinds of control issues Abraham found so vexing would rise to the surface. Being cared for by kind but personality-free nurses would only make anyone's hospital stay more inhuman and more emotionally restricted. "Hey," I had essentially just said, "I've got friends, and I'm not good at using a cell phone. I don't have a hole in my stomach, but, like you, I still have very human concerns." Discussing ordinary complaints and ordinary foibles could offer needed relief from the relentless seriousness of cancer.

Dr. Anderson showed up soon after this exchange, looked at Abraham's wound, and said it was too early for plastics to get involved. In person he wasn't quite as nice as I had hoped, but he struck me only as young and ambitious—the human explanation behind the driven surgeon stereotype. The nurses carefully re-dressed the wound, and Abraham was left with irritably waiting for his large incision to heal.

I spent some time over the next few days thinking about the window into my personal life that I had given Abraham. I began to wonder if the rules we learned in school about not revealing personal information were really designed to protect the patients from overly talkative nurses, or if they were meant to protect nurses from establishing an intimacy that would only make it more painful if the patient began to decline. Then I thought about my own life and how my being a nurse affects my kids. I use my job to teach them interesting things about physiology and medicine, but I also feel the need to protect them from the grim realities of my job, from the knowledge of sudden death and how long and difficult the course of treatment is for the majority of patients with cancer.

My husband gets no such protection. He's an adult after all, and in my first several months on the floor the need to talk about the day would be so overwhelming that I would blurt out stories almost without stopping to breathe, unaware, unless he told me, that I was so heavily buried in the lingo of the job that he often did not understand what I was telling him. Arthur is a theoretical astrophysicist, a cosmologist who studies the Big

Bang and the origin of the universe. During the years I stayed home with the kids, I often felt like nothing in my day could compare with the elegance and abstruseness of what he spent his day considering. It was odd for me to now have a job with its own impenetrable jargon, hard-to-get-your-head-around ideas, and elegant contemplation of what makes life worth living. However, regardless of his lack of understanding, I kept talking. I was confident that he could grasp the gist of my stories, if not all the details. Plus, I really needed someone to listen.

But the kids did not have a framework in place to make sense of my job or the issues that arose there, and I certainly did not expect them to be my emotional support. My husband was out of town when I had my first Condition A, and my mom had come in from Chicago to help me out at home. I got off at 3:30 that day and called her to explain what had happened. Her sage comment was "That really puts having a bad day in perspective." I gave a vague and brief explanation to the kids, then told them we were all going to our neighborhood Italian place to have a really nice dinner. I spent that dinner dazed from shock and drunk on the knowledge that all the people I cared about in the world were still alive. I watched the kids, seeing them in a new way, enjoying their aliveness. After the second Condition A, I remember only feeling exhausted, too tired to revel in the kids' aliveness or the aliveness of anyone. I wanted to crawl under a rock and only come out once I had a guarantee that everyone I knew would stay well and whole, at least for a little while.

I wrote about the first Condition A for the *New York Times,* and my article was published in the "Science Times" section a few months afterward. Much as I wanted to keep the details of that story from my kids, secrecy became impossible once the essay was in print. For one thing, the kids were proud of me. They took the paper to school and showed their teachers the article. One of my daughters cut it out and posted it on the bulletin board in her room. My editor at the *Times* had "fronted" the article, put it on the first page, without letting me know he was going to do that, and getting the paper that morning became a cause for family celebration: "Look, Mommy's on the first page of the 'Science Times.'"

I worried that learning about my patient's gruesome death would make the kids worry about their own lives, but to my surprise hearing these stories did not make the kids concerned for themselves; instead, it made them worry about me. They, I think, found their own rock to hide under briefly, but being kids they did not stay there waiting for a guarantee of wholeness to come to them, they went out and looked for it. That's when the questions started, just two, asked off and on, without preamble or context: "Mommy, are you going to get cancer?" and "Mommy, are you going to die?" And I, who pride myself on a kind of stubborn honesty, lied every single time any one of the three of them raised either of these questions.

"Mommy, are you going to die?"

"Oh, no," I would say, without even stopping to think about it. "Not until I'm very old."

And, "Mommy, are you going to get cancer?"

"No," I would say, full of certainty. "I am not going to get cancer."

Always the same questions, always the same answers, until at some point all three children had been reassured often enough that they felt they could stop asking, and did. A psychologist might say it's wrong to lie to children about death, and in general

I would agree, but relating my work to the lives of my children was a unique situation. They were being forced to confront questions of mortality to an off-scale degree almost daily. I imagine that when police officers leave for work, they don't tell their kids, "Well, today's the day I may get killed," because even though they might be telling the truth, the truth in that case would be cruel and gratuitous. Firefighters don't tell their children, "Today's the day I get burned up," and no one on the bomb squad says, "Today's the day I could be blown to bits." Arthur and I have been able to give our kids a feeling of emotional and financial security that is real. Shattering that sense of security for the sake of owning up to the hard truths I'd learned on the job would have been self-indulgent and profoundly unfair.

And so as much as possible I don't own up to those truths. I pick and choose which details from work to bring home to them. Taking Abraham as an example, I might tell them that while there's nothing wrong with prayer, for curing cancer, put your faith in modern medicine. I might also tell them a story about the skirmishes that took place between him and his night shift nurse whenever he put the sign on the door reading NO 4:00 A.M. VITALS. If I told them that story, I would try to make it funny, but I would also make a point about how patients lose control of their lives in the hospital, and the measures they take to get some of that control back. I might describe in exaggerated detail the particularity with which Abraham took his pills, while making it clear that I still found him wholly likable.

I would not tell my kids that Abraham had a giant open wound in his stomach. They don't need to know that, and it would scare them. As I've gotten more familiar with the job, I even try to spare my husband some of the gorier details about my work. Willing as he is to listen, Arthur would probably be happier not hearing about stomach wounds and emergency colostomies. Even I would rather not have seen a lot of the things my job has shown me, frightening truths about just how frail the human body can be. The difference is, I signed on for this, I get paid to confront these difficulties. My husband listens because he cares about me; I lie to my kids about my own vulnerabilities for the very same reason.

Some friends of ours like to play a game called "Two Truths and a Lie." It's a simple game: each person reveals two true things about herself and one lie. The people listening have to guess which of the three statements is not true. No one wins or loses; it's just fun, a semistructured way of getting to know people. I stumped them once by offering "My father used gasoline instead of lighter fluid to start a cookout," "The muffler fell off our car, and my dad fixed it with duct tape," and "My stepmother used to carry a gun in her purse." The lie you can figure out only by learning that whatever his proclivities for fire, or his attitudes toward self-protection, my father knows a thing or two about fixing cars.

The kids heard these three statements, mulled them over with me, and watched our friends as they struggled to figure out which statement was the lie. Some harsh truths about my dad's rural background came out in those simple sentences, but the kids have a context for understanding them, so the information, while possibly surprising, was also revealing. If I told them the story of Abraham's gaping abdomen and of his colostomy, they would have no context, no personal interest to mitigate their fear.

Also, what lie would I tell them? The biggest lie of all would be that of course we would cure Abraham's cancer. My kids are old enough and savvy enough to know that's not true.

So I parse these work situations the only way I can, with my own version of two truths and a lie. "Do people die at my job?" Yes. "Is cancer a horrible disease?" Yes. "Will *I* get cancer and die?" Absolutely not. My kids know that this, too, is a lie, but they need to hear me say it, to tell them that to the best of my ability I will protect them from the horrors I come across on the job by keeping myself intact and alive.

I protect them, but I've stopped trying to protect myself by keeping my life a secret from my patients. My brief and banal conversation with Abraham about cell phones and landlines showed me that keeping a lid on my personal life does not protect me anyway, it just makes everyone a little more uptight than they might be otherwise. So I talk about the kids, that there are three of them, that I have twins, their ages. If it comes up, I say I used to teach English. I talk about where I live and that when the weather's reasonable, and even sometimes when it's not, I ride my bike to work. It's small talk, really, but I get the feeling patients want to know these things to make me more human, so I tell them, bringing us all down to earth.

I never thought I would lie to my kids about anything. Arthur and I somehow kept up the fiction of Santa Claus without ever directly saying, "Of course he's real." We failed spectacularly at being the tooth fairy because we couldn't maintain the illusion. Now when one of the twins loses a tooth, she just asks us for the money up front, tooth in hand. We're facing the rush of puberty head on, and we try to be honest about the inevitable ups and downs of life. In talking about my job, though, the truth has no place. Matters of life and death are too serious to be cavalierly honest about with children. If my own death is imminent, if I do get cancer, then of course I will tell them the truth, but until then, we preserve the fiction that I am, relative to my job, invulnerable.

I learned not to open up to my patients, and instinctively I believe in always being honest with my kids. But my job flipped both those assumptions on their heads. If a conversational opening arises with a patient, I take it. It could be a chance for me to lead him back to his own humanity. With my children I slam the doors on my grittier work experiences and keep them locked. When I fell and hurt my knee, the kids learned, in a very real way, that Mom is not unbreakable. For now that's enough knowledge of life's random cruelties. I don't need to rub their noses in it by talking endlessly about my job. The harsh truths of life will find them soon enough.

EIGHT

Doctors Don't Do Poop

Nurses deal with slow death, constant sadness, gross inefficiency, and the chaos of failing lives, but we also deal a lot with patients' poop. And not just poop. We also deal with pee, mucus, sputum, vomit—all the bodily excretions—in their normal and blood-inflected forms. Typically, it's the nurse's job to measure the volume of these excretions, document what they look like, flush them down the toilet for patients who aren't able, wipe them off the floor when patients miss, and when necessary collect them, label them, and send them off to the lab.

This is the gross and vulgar part of nursing, and it doesn't get talked about much, except among nurses. In *The Things They Carried,* Tim O'Brien's novel-memoir about the Vietnam War, O'Brien explained the profanity in his book this way: "Send guys to war, they come home talking dirty." The same could be said of nursing: send nurses in to clean up poop, they'll end up talking a lot about shit.

One day I spent the entire day, every hour like clockwork, cleaning up a patient who was "incontinent of stool," which meant that every time he had a bowel movement, it came out more or less spontaneously, without his having any control. *Stool* is, of course, our grown-up medical word for poop, and stool can be "formed," "loose," or "liquid." This particular patient's stool was liquid and mucusy. It honestly looked more like something that would come out of someone's nose than their rear end, but come it did, and plenty often.

The patient, Mr. Barton, was an older man who'd been diagnosed with leukemia and then had his body wracked by chemo. We had tested him for *Clostridium difficile*, a hospital-acquired infection that gives patients constant liquid diarrhea, but he didn't have it. The doctors thought his diarrhea was caused by the chemo. Many kinds of chemo target all quickly dividing cells in the body. That means they kill cancer cells, but they also kill off cells in hair follicles (leaving patients bald), as well as cells in the gastrointestinal tract. As a result patients get nauseated, vomit, and have diarrhea. Usually with medication we can keep these symptoms somewhat under control, but nothing was working for Mr. Barton.

The aide and I spent the day taking turns cleaning him up. He was in isolation for VRE—vancomycin-resistant enterococci, another type of hospital-acquired infection—so going into his room meant putting on a yellow, paper, full-body gown, along with latex gloves. One of us would go into his room to clean him up, and thirty minutes later his call light would come on, and the other of us would sigh and head up the hallway. Mr. Barton's son was in the room, and he helped however he could, but getting Mr. Barton clean was really a two-person job, and someone with experience needed to be in charge.

You'd think that a day like that would come about as close to hell as it can in nursing, but I don't remember it that way. Even though his room was very small, crowded even with his son and the cot his son slept on during the night, and even though it gets hot pretty quickly under those isolation gowns and latex gloves, I don't remember it as a bad day. First, Mr. Barton himself was a nice man, and he accepted his incontinence with dignity. If it made him angry that he kept shitting all over himself, he didn't take it out on me or the aide. Whatever embarrassment he felt he kept to himself. His son, Josh, was also matter-of-fact. Neither of them talked much, but the support running between them was obvious, and it made it much easier for me to do my job.

But I still can't say that I found constantly cleaning up Mr. Barton a pleasant experience, and neither did Scott, the aide. Every time Mr. Barton's call light came on, Scott would look at me and say, "You're killin' me." Another aide, trying to be helpful, asked if we couldn't insert a "rectal trumpet" into Mr. Barton's rear end to collect his stool in a bag that would hang off the edge of his bed. However, sticking a semipermanent tube in someone's rectum brings its own set of problems—increased risk of infection foremost among them—and Mr. Barton wasn't so bad that he needed such a problematic solution.

The night shift nurse had told me the skin on Mr. Barton's bottom was so tender that "even a wet fart" would hurt, and this was not a joke or gratuitous scatological commentary, but, alas, the truth. In order to protect his skin, Scott and I had to keep Mr. Barton as dry as possible, and clean—a hard job when your patient's rectum is leaking stool. We would have him roll onto one side, wipe him, and then apply a moisturizer so thick and white and chalky it looked like human spackle. With Josh's help, Scott or I would then fold up the dirty pad beneath Mr. Barton, shove a new clean pad underneath the dirty pad, and have him roll to the other side, where we could pull out the dirty pad, smooth out the new one, and make sure his skin on that side was protected, too. Mr. Barton was really good at rolling, which made our job much easier. If a patient cannot roll, cleaning him up becomes a two- or even three-person job, and on a busy hospital floor, constant incontinence that needs three people for a clean-up means a lot of time off the floor not getting other stuff done. It's not that we don't want to clean up patients, we just don't always have the time to do it right when it needs to be done.

Around 10:00 A.M. I got a reprieve in the form of Mr. Barton being called for a CT scan. I pondered how best to handle sending him to CT. On the floor he was not wearing a diaper because we worried that it would lead to increased skin breakdown, but Mr. Barton himself had requested a diaper for his trip to CT. Being incontinent with all of us was one thing; pooping on himself in CT was quite another. I understood why he regarded the two situations as different, but I worried about his skin, too. I had no idea how long he might be down there sitting in a dirty diaper. I decided to ask Eleanor, an older African-American woman who had been one of my first preceptors. After I explained the situation, I asked her about using a diaper on Mr. Barton. "Oh, absolutely," she said, teaching me that it's important to protect our patients from the embarrassment of public incontinence, even if their skin suffers a little bit.

CT called and said they were ready for him, and I explained that I needed to clean him up and put a diaper on him. "He's been having this problem with frequent liquid stools," I said. "Oh great," the nurse in CT responded. She didn't want to deal with his incontinence any more than he wanted her to. Now I understood that wearing a diaper was in Mr. Barton's overall best interest: it would help preserve some small amount of his dignity, and it would make the people in CT more likely to treat him kindly.

Scott helped me get a diaper on him, and we sent Mr. Barton on his way with a couple of bags of wipes and some extra diapers, just in case. Changing patients is dirty, time-consuming work, and I didn't know if they had diapers in CT. If Mr. Barton had diapers with him, I thought it more likely they would change him down there rather than leaving him sitting in his own stool. It may sound heartless and even scandalous to say that, but it happens, because hospital staff are often too rushed to do the right thing, especially if "the right thing" means wiping someone else's bum, and that someone is only your patient for thirty minutes or so before he goes back to his own nurse on the floor.

An hour or so later Mr. Barton got back from CT, and I went into his room to help move him from the stretcher to the bed. He looked at me and said, "Now I know how a baby feels."

I looked back at him, confused. "What?"

Lying flat on the bed, he moved his legs up and down, knees bent, and said, "Waah—I want my diaper changed."

"Oh," I said, suddenly understanding. Then I realized what a stroke of genius his baby imitation had been. Wearing a diaper made him feel infantilized, I'm sure, and so he took that feeling of infantilization and acted it out as a joke to get what he needed. He was at least six feet tall and took up a lot of space in that crowded room. The incongruity of this tall, gray-haired man kicking his legs like a baby and whining "Waah" got the message across in a clear but playful way: please clean me up so I don't have to sit in my own shit any longer.

So I cleaned him up. It was actually a nice moment. In general, he and his son were both reserved; I think that worry made them quiet. But Mr. Barton's baby act broke down that reserve, if only briefly. Suddenly, it became possible for him to respond in a way other than patient stoicism, and I was then able to relax a little, to chat a bit, to be personable rather than merely kind and efficient.

The diarrhea went on and on, until some point later in the day when Mr. Barton's stool became blood-tinged, even more "mucusy," and liquid. Earlier in the day it had looked like something that might drip from a person's nose if he had the worst cold ever. Now, however, it looked alien, scary, more like an abortion than a secretion. The bright yellow color had disappeared, and Mr. Barton's stool had become a gelatinous admixture of clear mucus and bright red blood. It was a fluid that bespoke illness and violations of the natural order. It concerned me, and I decided to call the intern and let him know about this change in the products of Mr. Barton's GI tract.

Of course, the intern wanted a stool sample, so the next time Mr. Barton put his call light on, I went in to collect it. We had been having some luck getting him to use a bedpan. Mr. Barton's son would put the bedpan under his father when he needed it, then call us in if Mr. Barton ended up using it. The bedpan helped with the mess and was somewhat better for Mr. Barton's skin, although he didn't always get everything into the bedpan, and sometimes his son just wasn't quick enough. This time, though, the bedpan had worked, and I took it out from underneath Mr. Barton and into the bathroom, where I poured the contents into the clear plastic container we used for stool and urine samples. Technically, I suppose, I did have a "stool" sample, but the bloody and viscous sludge in the orange-topped jar looked like no stool I had ever seen in my life.

I printed off the lab requisition I needed and appropriately labeled the jar. The nurse who had Mr. Barton the day before stopped and looked at the sample I had collected. "His stool definitely did not look like that yesterday," she said.

At that moment the intern himself came walking down the hall. I could not resist showing him Mr. Barton's stool sample. I held up the plastic container and told him what it was. He looked at it and said, "Ugh, that's disgusting." I looked back at him. "Hey, buddy," I wanted to say, "you didn't have to pour it into the container. And you sure haven't spent the day cleaning it up." It's interesting, though, that even doctors can be squeamish about shit.

Doctors don't do poop, but nurses do poop up, down, and sideways. We have confused patients who stand up and poop on the floor, patients too ill even to know when they poop, patients who will save their poop for the nurse, put on the call light, and insist that I look at it *every single time* before they flush. I have never seen a doctor on rounds even gesture toward the idea of cleaning up a patient sitting in her own stool or collecting a needed stool sample. I can say with almost 100 percent certainty that no doctor in the United States has ever collected for herself the sample needed to check for "occult," or nonvisible, blood in someone's feces. The kit for doing that comes with a balsa wood tool that looks disturbingly like a coffee stirrer, and you need "stool" from two different areas of the poop to get a good sample. When the waste is solid, it's easy enough, but if your patient's stool is viscous and swampy, well, all I can say is, poop on a stick: even for nurses, it's harder than it looks.

It creates a strange kind of community, all this dealing with poop. "I'm really in the shit today," people will say, and we all know it's meant literally. Poop gets described and classified by its color, consistency, constancy, and smell. For a while there, one aide updated us on which patient had the "stinkiest poo." Nurses will talk about diarrhea over lunch, dinner, even drinks. The morning I had Mr. Barton, I charted while eating a granola bar. I had just taken a bite and was chewing while checking off the appropriate boxes in the medical record to describe his stool. I must have had a sour look on my face as I considered the fine distinctions between "loose" and "liquid" because an aide walked by and said, "Theresa, either that's a really bad granola bar or you're having a problem." "Oh," I said, "I'm just figuring out how to classify my patient's stool," never thinking about how incongruent it was to be eating while I did this.

Shit stinks, and there are evolutionary reasons why we instinctively shy away from human feces. The *Escherichia coli* bacteria that line our intestinal tract can make us very sick if they get swallowed, which can happen if people don't wash up after going to the bathroom. Simply telling another nurse about a patient's "liquid green stool" can make me get up and wash my own hands, very slowly, with lots of soap and water. But when people get sick, really sick, they often lose control of their bowels, and someone has to clean them up, keep track of how much fluid they are losing, moisturize the skin on their rear end, and tell them, over and over again, that it's really OK, that we don't mind cleaning them up at all.

That was my day with Mr. Barton, busy but predictable, until his son saw a sample of what Mr. Barton's stool had turned into. I had thought he was aware, but in the late afternoon I pulled out the bedpan, and Josh got a good look at it for the first time in a few hours. Seeing the blood, and the swirling, transparent sludge that seemed vaguely placental, Josh got upset. He'd had no idea his father's stool looked like that—so different from normal feces it didn't appear human—and he was very concerned about the blood.

I didn't have a good answer for him, so I once again paged the intern. Nurses are trained to be patient educators, but I happily punted this question to the doctors. I had no idea what was going on in Mr. Barton's gut and did not want to speculate.

When the intern paged me back, I explained Josh's concerns and asked him to come by to talk to the two of them. The intern was in the middle of three different things and couldn't come right away, or even soon, so he kicked the ball back to me and left me having to answer Josh as best I could. The core of this explanation centered on the blood, how a small amount of blood can make even a large amount of fluid look bloody. That's true, but Josh did not find what I said too comforting. How could he? Human waste should not look like a jellyfish that has been pureed in a blender, and it's understandably distressing when it does.

A day and a night can be a long time in the hospital, and when I returned the next day, Mr. Barton's stools had become much more normal. He was no longer incontinent, and the bright yellow color had returned to his stool, along with a more normal consistency. The blood was gone. Scott, the aide, got the idea of putting a commode in the room right at the bedside and moved Mr. Barton to a chair so that transferring him back and forth from the chair to the commode would be relatively easy. I made many fewer trips in to see Mr. Barton that day, and each time I did he had improved remarkably. By the end of my shift that day, he looked like a different person from the man I had taken care of the day before. His face was much more relaxed and made him look his seventy-odd years, instead of much older. He sat up straight, no longer infantilized by his own body, and talked more. Because the rhythm of his life was no longer ruled by shit, he could much more expansively be a human being.

After that shift I had a few days off and was delighted to learn, when I next came to work, that Mr. Barton had been well enough to go home. His bowels had gotten more under control each day, and as they calmed, his energy grew, and his resilience returned.

We poop to eliminate waste from our bodies. Liquid stool, blood in our feces, incontinence: each of these symptoms means that some process in our body is out of whack. If nothing is done to restore the body's balance, people die. Indeed, death from diarrhea is a reality in economically disadvantaged nations that lack sufficient health care. Cholera kills not because the disease is so virulent, but because it causes deadly dehydration that only IV fluids can control. So, I empty bedpans, I study what's in them, sometimes I pour the contents into a plastic container and send it to the lab. It's gross, but it matters, and it's part of what nurses do.

Several months passed before Mr. Barton was my patient again. I was working from three to eleven P.M., and I picked him up when I came in. Mr. Barton had not been having any digestive problems for a while now, but his cancer had returned. I learned all this very quickly in report, and heard also that the attending doctor was going to talk with Mr. Barton and his son that afternoon. I had no idea what the content of this "talk" would be, but knowing that the attending had scheduled a family talk for 4:00 P.M.—when attending physicians are almost never on the floor—told me the conversation would be serious.

The doctor arrived right on time, and he, the fellow, the intern, and I all trooped into Mr. Barton's room. This particular attending, Dr. Priam, is a big man who is completely bald, and an émigré. People like to describe him as a giant teddy bear, but he has an imposing side that can be physically and intellectually intimidating. He was not someone I would address directly if I could possibly avoid it, and I knew that my role in this "conversation" would be as a listener only. Still, I wanted to know what was said and how the family responded.

We all stood in the room while Mr. Barton and his son sat. I always think this is a terrible way to conduct these conversations—those of us with the knowledge and the power tower over the folks who have to hear the bad news—but the truth is, hospital rooms tend to be small and typically do not have enough chairs to seat extra people, even for discussions of life and death. I also wondered if we all needed to be there in such a large group, but if we all weren't in Mr. Barton's room, we wouldn't know what was said. The nature of a teaching hospital is that all of us were involved in this patient's care, just with our responsibilities defined in terms of a rigid hierarchy.

Dr. Priam did not mince words. I was there for the entire conversation, but it felt as if I had missed half of it since he seemed to start in the middle of a discussion that had been going on for a while. "Your disease has returned, and we cannot do another round of chemo because of the toxicities from last time," he said, in his thickly accented but precise English. The situation Dr. Priam described was not that unusual in cancer treatment. Chemotherapy drugs are toxic, potentially deadly poisons, not just for the GI tract, but for other major organs as well. Some chemos can damage the heart, others the liver, and some the kidneys. The lungs can suffer and the nervous system, too, leaving patients with permanent numbness, tingling, and pain in their arms and legs. Giving patients chemo always involves balancing risks and benefits. The benefit can be saving someone's life, but for some patients, and Mr. Barton now belonged in this group, the risk is death.

"We have three options," Dr. Priam said, holding up his large hand and extending the first three fingers. "One," he explained, extending his forefinger, "we can give supportive care and also try to make you as comfortable as possible." The son, Josh, frowned when he heard this, and Mr. Barton looked puzzled. "Two," Dr. Priam said, "we cannot give any more treatment and only make you comfortable." Josh frowned deeply when he heard this and shook his head, looking down at his lap. "Or three," Dr. Priam said, and he seemed to sigh just slightly, "we can give some chemo, but it won't be enough to cure the disease, and it will have some of the same toxicities as before."

Josh nodded at this and then asked, "But what about his ankle?" Mr. Barton had some cellulitis, a skin infection, on his right ankle. He had been admitted because of it, and we had cleared it up pretty well but had not been able to get rid of it completely. "What about the treatments for his ankle?" Josh asked.

Dr. Priam waved his big hand. "You are focusing on the wrong thing," he said. "The ankle is not an issue—the issue is what you want to do about treatment."

Denial may be the most powerful psychological force in the known world, and I admired Dr. Priam for his ability to confront Josh's denial so unsparingly. It is very, very hard to tell someone he is going to die, and it can be even harder to tell his loved ones. Dr. Priam's manner would probably not be described as therapeutic, but it was honest in a situation where the patient and his son did not want to face the truth. Once he finished laying out the different choices available, Dr. Priam asked Josh and Mr.

Barton if they had any questions, and they both said no. Josh looked determined, and Mr. Barton looked bewildered, as if he hadn't understood. I knew he was mentally intact, but his look of bewilderment was so intense I wondered if I needed to reassess his mental status. That's how it can feel to look death in the eye—the desire to look away, to not listen, can be so strong that it makes people seem deaf and blind.

We all trooped out, and on my way I looked inquiringly at Josh. Without speaking, I was trying to say, "Do you need anything? Can I help?" He was still looking down, frowning, and his father remained speechless and confused. Seeing how stunned they were made me realize the futility of empathy or even speech. There were no words to make Josh's or Mr. Barton's pain go away, and however much I listened or cared, the pain would be theirs to bear, not mine.

Later, though, I came back in to check on them, and Josh told me his father wanted a Bible. He had meant to bring one but forgot. A couple of other family members had gathered by this time, and they made motions to get a Bible themselves, but Josh said, "Theresa will find him one."

I'm not a religious person, but nothing on that shift became more important to me than finding Mr. Barton a Bible. I asked our secretary where one might be and scoured our supply room. When none turned up there, I went over to our sister unit and asked them if they had any Bibles. They said I could look in their supply room, but that I also might want to try the chapel on the first floor. Their secretary said the chapel was always open and that it was supposed to have extra Bibles for patients. I searched their supply room, but it also contained no Bibles. I marveled at what a secular society we had become—how could it be so hard to find a Bible in a hospital? Undeterred, I took the elevator down to the first floor and went to the chapel. The door was unlocked, and there were a few random Bibles on a shelf near the door. No signs indicated that these Bibles were there for the taking, but there was also no one to ask, so I took one. Since Mr. Barton was older and wore glasses, I picked the one with the biggest type, not having any other criteria for selection.

We were coming up on 11:00 P.M., the end of shift, so I hustled myself back up the elevator to the floor. "Hey, look what I found," I told Harry, the night shift nurse who had passed Mr. Barton off to me three months before with complaints of pain following wet farts.

Harry has a great chuckling laugh, and he laughed now. "What's that?"

I held it up. "It's a Bible," I said triumphantly. Then I lowered my voice, "Dr. Priam had 'the talk' with them today, and they wanted a Bible." We both shook our heads. Such a small offering in the face of such unquestionably bad news, yet it was the only thing they had asked for.

I took the Bible into Mr. Barton's room. Josh looked up and said, "Ah, Theresa found one." I handed it over wordlessly. Josh and his father had already told me they did not want to talk about what Dr. Priam had told them, that they wanted some time to think. I gave Mr. Barton the Bible, nodded, and left.

Mr. Barton was never again my patient, but I heard through the grapevine that he had pursued some combination of treatment and palliative care. I didn't hear that he died, but I'm sure he did. Sometimes it's easier not to ask, to let my own denial keep me happy for as long as it can.

In *The Things They Carried,* Tim O'Brien explains that a "true war story cannot be believed" because "often the crazy stuff is true, and the normal stuff isn't, because the normal stuff is necessary to make you believe the truly incredible craziness." When I first started working as a nurse, the hospital seemed like the least "normal" place I had ever been. We stick tubes in every possible human orifice, slice people open to save their lives, fill their veins with poison, measure their urine, count their bowel movements. The craziness *is* normal, and the only thing that's really normal is the fundamental humanness that unites us all. Sometimes a patient needs his bum wiped twelve times in half as many hours, and sometimes he needs a Bible. Soup to nuts; shit to death—we're all on the same continuum.

Anyone hearing a true nursing story will not want to believe it. The level of vulnerability, dependence, and fear experienced by patients in the hospital remains far outside the realm of normal, everyday life, and none of us want to imagine ourselves in that position. But people find themselves there, regardless, and they find nurses there, too. Doctors don't do poop; they're concerned with other things. That's OK, but it's a difference between the two jobs. Probably they don't do Bibles, either. But nurses have to get to the heart of the matter, whatever that may be.

NINE

Switch

The floor is your life, your home, and, depending on how the shift is going, your prison. Every floor has its own character, and about seven months into my first year I realized my floor was feeling more and more like a prison and less and less like a home. By inclination I'm a loyal person, and I don't give up on people or institutions easily, so it took me a while to admit that I needed to change jobs—meaning that I needed to change floors—but the impetus, the moral urgency even, behind my decision became clear one night at work following one long shift.

That day one of my patients, Irene, found out that her lung cancer was in its final stages. Her family found out at the same time, and neither she nor her family had any idea her situation was so dire. The doctors had not adequately prepared any of them for the bad news, but partly no one knew just how badly she was doing because she looked so great, and until recently had been feeling pretty good. In her late sixties with adult children and grandchildren, Irene had gained local fame as a talk show host, and despite her disease she remained a beautiful and vibrant woman who looked much younger than she was.

She'd had a chest tube inserted into the space between the two layers of tissue in her lungs—the pleural space—to drain fluid that was accumulating and making it hard for her to breathe, but the chest tube had been pulled the day before, an acknowledgment that no matter how much fluid was drained out of her lungs, nothing could really be done to improve her respiratory status or prolong her life. She had confusion that came and went, pain in her lungs, and episodes of air hunger, where she would gasp for air despite her oxygen saturation reading within normal limits. As a nurse, I find watching a patient panting for air very difficult; for family members, the experience is agonizing. Morphine is the usual treatment because it tricks the brain's oxygen receptors out of panic mode, giving the patient some relief from that feeling of suffocation. Of course, morphine would also increase Irene's confusion, making it harder for her to know her own mind, understand her grim prognosis, and communicate her wishes to her family.

These are all the things I was juggling: her pain, her difficulty breathing, her children's shock at discovering their mother was really dying, and the patient's own waxing and waning grip on awareness, and even on control over her own body and bodily functions. There were two daughters in the room and a daughter-in-law, but the one son ended up being the manager of the situation. Over six feet tall, he had inherited his mother's good looks and her youthful appearance. He was gracious and direct, and I worried that due to my own relative inexperience, I would not be able to give him or his mother the thoughtful care that the situation required. However, I intended to do my best.

Irene maintained her composure and kept her dignity even when she needed help to and from the bedside commode or when she couldn't breathe. With her thick jet-black hair, pale skin, and perfectly arched eyebrows, she looked like a kinder, gentler version of the evil queen in *Snow White,* or maybe just a middle-aged version of Snow White herself. She didn't say much, because she had a hard time choosing words, but she had presence, and that presence put her at the center of all the concerns and questions swirling around the room.

My first clue to the difficulties this patient would present came not from the night shift nurse or her doctor, but from the patient care coordinator, or PCC, the nurse who arranged care for patients following a discharge. It would be nice, certainly beneficial, to have clearer channels of communication regarding patients' status and disease progression, but in the controlled chaos of the modern hospital, a nurse is thankful for any source of accurate information. The PCC, a thin, wiry woman who looked, in contrast to Irene, all of her sixty-plus years and wore a constant look of kind determination, told me in a whisper how outraged she felt that Irene's family had no idea how sick Irene really was.

Later in the day one of the daughters informed me, also in a whisper, that Irene herself became aware of her poor prognosis that day in the hospital while watching the TV channel where she normally appeared. A reporter had announced to their collective television audience that Irene was dying: it was the first she had heard of it.

I probably should have been outraged, too, but I was too busy trying to figure out what was actually going on and what I could do to help. In the meantime, I had three other patients, all of whom needed meds and one of whom needed chemo at the same time in the late morning when Irene's surgeon, a big-framed and charming African American with a surprisingly cherubic face, came to talk to the family about Irene. I felt overjoyed to be included by the surgeon in his discussion with the family, then deflated when I had to excuse myself to hang chemo in the room next door. Hospital nursing often involves choices like this: in caring for the patient who needed chemo, I missed out on the surgeon's discussion of Irene's status, and on the family's response to his news.

Still, the pieces slowly came together, and my priorities didn't change: Irene's pain, her troubled breathing, her poor prognosis, the decisions that needed to be made by her family. Early in the afternoon the attending doctor had a meeting with the family, and again, I was too inexperienced to understand that I was entitled to be there, that I didn't need to wait for an invitation, so I missed that meeting also. This time the PCC updated me: the family had decided on hospice, and Irene was going to be put on a PCA pump (patient-controlled analgesia) so that narcotics could be used to control her pain and help her breathing.

At that point in my career, I had never hooked a patient up to a PCA on my own. I had a vague idea of needing to order the special pump from central supply, and I knew the glass cylinders filled with morphine and Dilaudid were locked up in the Accudose machine, where all our narcotics were kept. I was just starting to wrap my mind around what I needed and whom I could ask for help when, in a gracious and controlled way, all hell broke loose in Irene's family. I had missed the family meeting, but key members of the family had missed it as well and felt they didn't have all the information they needed to consent to hospice care for Irene. Even more important, Irene herself had suddenly become more lucid, and although she wasn't refusing hospice, she needed to have the doctor tell her personally that she was dying, and she needed him to tell her at a time when she would be able to understand.

Once again Miriam, the wiry and kind PCC, came to my rescue. She explained that I needed to call the attending and explain the family's decision not to continue treatment, but to postpone *stopping* treatment, which included putting the PCA on hold. It was an unusual situation, but I could imagine that if I had learned of my own imminent death while watching TV in my hospital room, I might want a face-to-face conversation with my doctor about it. The family wondered if the doctor could come see Irene right then, since she was aware and able to talk. It fell to me to communicate that wish by asking the doctor to return.

Wow. Call the attending, no, *page* the attending and ask him to come back to Irene's room. Paging attendings was not part of my nursing repertoire. Approaching the residents and interns didn't faze me at all—most of them were about half my age and reminded me of my former Tufts University students. But an attending was different. The attendings occupied a separate clinical world: they appeared for rounds, accompanied always by their entourage of new doctors, fellows, pharmacists, and sometimes nurse practitioners and physician's assistants. I had seen the residents apologize when they asked the attendings for things: "I'm really sorry, Dr. So-and-so, but can you sign this narcotic scrip for me?" The attendings did not bustle around looking stressed the way the doctors-in-training did. They spent rounds listening, probing, guiding, chastising, sometimes with an air of great impatience, then disappeared to handle more important matters than the mundane this and that of daily inpatient care.

This particular attending, Dr. Rohe, did seem very nice, avuncular even, but I still felt very nervous putting in the page and waiting for his return call. When the call came, I found myself almost tongue-tied in my effort to be ingratiating. "Um, even though you spent a long time talking to the family, they feel like they didn't completely understand what the options are for their mother," I told him, wincing to myself. "They're wondering," I continued, feeling as if I might as well get it all out at once, "if you could come back and talk to them all again today."

My earnest, breathless speech ended, and for a very brief period, during which I braced myself for a huge telling off, the doctor said nothing. Then he answered, in the gentlest voice I could have hoped for, "Well, I'm on my way now to pick my daughter up from school. How about if I talk to them tomorrow morning?" He was on his way to pick his daughter up from school. Dr. Rohe was a human being after all. His tone was even slightly apologetic, as if he regretted that his conversation with the family had left some of them confused.

At this moment something clicked for me as a nurse, and I understood that my role was not to be a passive helpmate, or a diffident flatterer, but an involved partner. I wondered what to do about the PCA since the family wanted to wait on it, and I asked Dr. Rohe. He, in turn, asked me, "Can we just put that on hold until tomorrow?" "Sure," I said. Even though I had never even set up a PCA, it made sense that it could be put on hold temporarily if that was what the doctor wanted. "Can I put in a free text order to hold the PCA pending your conversation with the family tomorrow?" I asked. Sure, that was fine with Dr. Rohe.

I finished the conversation with him, then went to tell the family the result. Irene's son and daughters looked at me with tense faces. "Dr. Rohe asks if he can talk to you tomorrow morning," I told them, feeling tentative. "He's on his way right now to pick up his daughter from school." As soon as I explained, Irene's family relaxed. "Oh well," they agreed, "if he's picking up his daughter from school." We all seemed comforted by this very ordinary and yet important commitment of Dr. Rohe's. Who could argue with a daughter needing a ride home from school? I explained that we would wait on the PCA until they all talked the next morning. I could keep Irene's pain and her breathing under control with injections of IV narcotics. Irene still wasn't able to do much talking for herself, so the family spoke for her. They didn't want her too sedated in the morning when Dr. Rohe arrived to talk. I promised to outline the situation for the night shift nurse, stressing Irene's desire to understand her prognosis, and explaining how her changing mental status complicated that understanding.

By this time it was close to the end of shift. I had taped report for the night nurse and finished up what needed to be done for my other three patients. I put the order about holding the PCA into the computer and went back to see if Irene or her family had any other concerns.

It was now 7:30 P.M., and my shift was officially over. I could have, in good conscience, left the floor and gone home, except that the nurse taking over for me usually showed up late and started her shift late. I did not want to leave Irene without a nurse on the floor, and I knew that Crystal, the nurse coming on after me, would spend the time from 7:30 to 8:00 P.M. hidden in a corner of the floor scouring her patients' electronic records to make sure that all t's had been crossed and all i's dotted. Some nights that would have been OK, but this night, at about 7:40 P.M., Irene suddenly had a much harder time breathing.

I still remember how she looked, sitting up in bed, her black hair in place, her beautiful face trying to look calm, but she was panting for air. Right away I turned up her oxygen to see if that gave her some relief. If someone is gasping for oxygen, giving her more can be a good short-term solution to help her stabilize. The oxygen didn't help, though, and indeed her oxygen saturation was in the mid- to low 90s—not ideal, but within normal limits for someone with her respiratory problems. The family asked how she could be open-mouthed, straining for air, when her oxygen levels were normal. "I'm not sure," I told them, because even though I knew it happened, I hadn't thought through the physiology of air hunger. Irene could have been collecting carbon dioxide, and the resulting acidosis would make her desperate for air. Anxiety and breathlessness can set up a vicious feedback loop where slight shortness of breath triggers an attack of anxiety, which then increases the patient's shortness of breath, which leads to more anxiety, and air hunger can be the end result. Whatever was causing Irene's confusion might also have interfered with the oxygen receptors in her brain, telling her that she was oxygen starved even though she wasn't. When patients are close to death, as she was, key bodily functions slow down. Often it's the kidneys, so that patients near the end of their lives stop producing urine. Irene was still peeing, but her lungs may have been starting to give out.

Regardless of the reason for Irene's air hunger, I needed to do something to relieve her desperation. She really was having a hard time, and I called over a couple of other nurses to ask their advice. The quickest fix was morphine, but her family was concerned that morphine would once again increase her confusion. She had breathing treatments ordered, and getting a dose of albuterol, a bronchodilator used to help asthmatics, through a nebulizer could help. I desperately paged respiratory and worked on convincing the family that a small dose of morphine would be a good idea in this situation. Because Irene's oxygen sats were stable, I wasn't worried she would crash, but it was also obvious that she was suffering.

I gave her a small dose of morphine, and respiratory showed up to give a breathing treatment. Pretty soon after she got the dose of morphine and the respiratory treatment, Irene's gasping slowed and then stopped, and she began breathing normally. I had averted the crisis, but it had been scary. Without the advice of other nurses on the floor, I wouldn't have known what to do. Certainly, without a nurse at the ready, things with Irene might have gone very badly.

It was now about 8:10, which meant I had overstayed my shift by forty minutes, when Crystal, finished with her preparations, came onto the floor and made it clear she had questions for me. Crystal is a tall brunette who kept her long hair pulled back into a tight ponytail. At times her face will break into an endearing childlike smile, and she can be expansive and helpful. More often, though, she inhabited the floor like a roving bad mood. She barked at people and saw it as her role to rigidly enforce policies and procedures, no matter how irrelevant they might be in specific situations. She had the role of clinician, which meant she was a floor nurse with some managerial responsibilities. She and the floor's other nurse clinician saw themselves as disciplinarians, people who knew the rules and made sure they were followed. But when it got right down to it, they were often just bullies. Both of them seemed to enjoy telling people off, and they rarely offered help when it was needed. Their criticisms and rebukes most often fell on the newest and most inexperienced nurses. That night I got my first real taste of what such bullying was like.

The questions Crystal raised about Irene's care, and my answers to those questions, don't matter nearly as much as the flavor of our exchange. This was a full-scale, ten-minute dressing down at full volume, and it was humiliating and enraging. The issue of the PCA she addressed as "Well, you can't just ignore an order!" On the family's request that Irene not be given any morphine from 6:00 A.M. on, Crystal said, "Well, I treat patients, not families, and if the patient wants morphine, I'm giving it to her." "No," I said, "the patient does not want the morphine, either." She went on and on, asking questions about Irene's paperwork to confirm the mistakes she already knew I had made.

What bothered me the most in this experience, even beyond the attack on my competence and Crystal's aggressive delivery, was the nurse sitting just two feet away from me at a computer in the same room. I knew this nurse and liked working with her. She was young and bubbly, extremely competent, and usually very fair. But through Crystal's entire tirade she said nothing, never even looked up from her computer. I did not like that. Bullying thrives only because people collude with it, and this nurse's silence appeared to me as a gross ugliness. I did not want to work with people who would let a nurse be so cruel to one of our own, and on this floor it happened over and over again. Public humiliation by senior staff, and especially the two clinicians, was so common it was considered normal.

Bad as this incident was at the time, I realized the most horrible thing about it only later. Irene stopped being my patient at 7:30 that evening, but I stayed and took care of her for another forty minutes while Crystal combed through her chart, looking for all the elisions and inconsistencies that she could then throw back at me. What if I had left? I wondered. What if I had walked out, as I was entitled to do, and Irene had been in her room, with her loving and concerned family, desperate for air, with no nurse available to page respiratory and to suggest and quickly draw up a dose of morphine? What if? I don't like to think about it, because the thought makes me very angry.

My husband finally picked me up that night at 8:30, and I was in tears. "If we can't help people die, what good are we?" I asked him, because lost in Crystal's rant was the very short amount of time Irene had left on earth. She was dying, and would die, in that same room, in just a few days. Her family was struggling to digest the news that she had run out of options. Helping them through that struggle, indeed, helping any patient in immediate need, would always be more important to me than paperwork, and no amount of bullying by Crystal or anyone else would convince me that putting patient needs before paperwork was wrong.

I talked to my boss, and I talked to the advanced practice nurse on the floor about this incident and others, but it became clear that, although my unit manager sympathized with my feelings and also saw Crystal as a negative influence, nothing was going to change anytime soon. More and more I came to dread going to work. I kept to myself on the floor as much as I could, but some ugly pattern had been put in play, and I was unwilling to, as a concerned older nurse had advised me, "be more submissive."

I knew I had to leave, but how, and where? Literally at the end of the hall and through a set of double doors that always stayed closed, my floor had a sister unit. This floor, which had a separate nursing staff and its own unit director and clinicians, was a medical oncology floor, but they also treated patients who had stem cell transplants, or what used to be bone marrow transplants.

I knew very little about the floor, but I did know the UD (unit director). This woman, Mona, who was tall and blonde where my manager was brunette and petite, had years of experience in oncology, and I appreciated her because after the Condition A when my patient bled out and died in seconds in front of me, she made a trip over to our floor to make sure I was OK. She had had a similar experience in her twenties, she told me, and it took her several years to get over it, a piece of wisdom I packed carefully away and still find useful.

One day I found out that Mona had also worked on my floor and had transferred to the sister unit after getting fed up with the floor I was on. She had recently been made director of the sister unit after working as a staff nurse and as a clinician. This nugget of information gave me the idea I needed for how I could change jobs and where I could go. I wanted to stay in oncology, and I did not want to switch to a surgical oncology unit. Could I solve my work problems, I wondered, simply by moving across the hall?

Of course, nothing is simple when navigating corporate hospital bureaucracy. I talked with Mona, and I talked with Diana, my manager at the time. Then one night after another horrible ending to a day at work, I ran out of patience and just applied online for a few different jobs at my hospital. I applied to our sister unit, and I applied to the emergency department and the ICU, mostly to show the powers-that-be that I was serious about leaving. I e-mailed Mona, and I told Diana, and then I waited, telling no one on my floor about my decision. Mona interviewed me, and I shadowed on the floor one Sunday morning. Right away I noticed a startling difference between the two floors. On this floor, no one yelled at anyone. Never. The tension that hovered in the air like a miasma was absent. It was still high-stakes inpatient health care, so the nurses had stress, tempers might flare, and more would need to get done at times than possibly could be, but there was none of the meanness and just plain spite that was inescapable in my present job.

A couple of weeks passed, then one day an HR rep called me at work to offer me the job. I said I wanted to sleep on it, hung up the phone, and realized I felt really happy for the first time in months. I could have called her back right away and accepted, but instead I examined my reaction to make sure it was real. Then I just let myself enjoy that feeling of relief and release for the few short hours left in my shift. By the end of my eight-hour day, I had told Mona and Diana. I called back the woman at HR and, in the formal language spoken only by bureaucrats, "accepted the position."

I told my friends on the floor that I was moving across the hall, and though some of them were disappointed, no one was surprised; that's how obviously bad it had been for me. I volunteered to finish out the schedule that had just come out because I didn't want to leave them shorthanded. I have to admit, too, to a flavor of defiance in my offer to stay longer than I had to. The clinicians had talked me up as having a "bad attitude," and I wanted to leave no opportunity for anyone to say I had screwed them. Probably I should have just said good-bye and been done with it, but I couldn't. I came to work every day I was scheduled, even worked extra. Maybe I also needed to prove to myself that it wasn't my fault things hadn't worked out, and even that I deserved this transfer since they let me switch jobs earlier than nurses are officially allowed to.

On one of my last days on the floor, the sheer impossibility of staying became completely clear. As usual it was a busy day, and

I was finishing up a patient's discharge papers. If the patient has a long list of medications, it can take some time to get the discharge paperwork together, and then the list of drugs has to be double-checked with another nurse. Usually the patient has been told about the discharge long before the doctor puts the necessary discharge information into the computer. As a result, by the time the nurse completes the paperwork, the patient is chomping at the bit. If there are hang-ups with insurance, questions about medications, or details still to be worked out for home care, or even if the patient has a long ride home, his impatience will only increase, often exponentially.

That afternoon I was finishing up discharge instructions for an impatient patient, and I had already been interrupted three times. I came back to the computer after the third interruption determined to finish the paperwork and get my patient out of the hospital. The floor had two long parallel halls with a nurses' station filled with computers in the center between the two halls. A bit down the hall I saw Anna, our other clinician, talking to another nurse on the floor. Then Anna saw me and said to that nurse, "Oh, that's OK, Theresa can help me," and said to me, really as more of a command than a question, "Can you help me do . . ." whatever it was. "Well, actually, I have to finish these discharge instructions," I told her.

It was probably embarrassing for her to be turned down in front of the doctors and other nurses sitting at the computers in the nurses' station, or maybe it just annoyed her that I said no, but standing there in the middle of the hallway, she snapped, loud enough for everyone to hear, "That's the problem with this floor—no one is willing to help you when you need it." I felt stung. After all, the nurse she had been talking to, who was a friend of hers, had agreed to help her. Why had she decided to ask me instead? It was an ugliness about the floor, this hierarchy based on power and connections.

But nursing is too difficult and too important a job for help to come with a hierarchy. Nurses should help where they're needed. I had endured on the floor mostly in silence, embodying some timeworn notions of gentility and honor I had absorbed during my southern Missouri upbringing. Not this time, though. I returned to my computer flushed and angry. Everyone sitting at the nurses' station had heard what Anna said, and everyone knew she was talking about me. I stood up and said loudly, "That's why I'm transferring floors," and then sat back down again to finish my patient's discharge.

Later that day and over the next few days, several of the interns and residents came to me and said, "Theresa, I heard you're transferring." These young doctors had been on the receiving end of Crystal's and Anna's hot tempers also, but I thought they just dismissed it out of hand. After all, they were doctors. However, when I said, "Yes, I told HR I won't work where staff yell at and humiliate other staff," and "I can't work with people who would rather criticize others than do their jobs," they nodded sympathetically, knowingly. The relief I felt at learning even the doctors found the atmosphere on my floor deeply challenging was tremendous. It dispirited me to discover that some of our nurses enjoyed pushing around the younger doctors—especially since those doctors needed our help—but as my transfer got closer and closer, it helped to know that people besides me found the floor difficult.

The time did pass, my last day eventually came, and I cleaned out my locker and said my good-byes. A week later, when we got back from our family vacation, I went one afternoon at three to the same hospital, went up the same elevator, got off on the same floor, but turned right, not left, and went through the closed double doors to my new job. The doctors were the same, the pharmacists unchanged; I even knew many of the patients. But the work environment was . . . normal. If I had a question, I asked it, and another nurse would tell me the answer without suggesting that only an idiot wouldn't know *that*. If I needed help, someone would offer without sighing and rolling her eyes. If I couldn't find something in the supply room, a more senior nurse would come into the room with me and point out, without sarcasm or condescension, where that dressing, or syringe, or bag of IV fluid was located. The nurses were professional, appropriate—overused words, but every day I went to my new floor, I marveled at how central those qualities were to doing the job well.

There's a phrase in nursing that applies in this situation, and though mentioning it feels like airing our dirty laundry, I'm mentioning it anyway: "Nurses eat their young." The first time I heard it, I wasn't even in nursing school; I was volunteering at the hospital where I had my twins, and one of the midwives told me, "You know, nurses eat their young, but not on this floor." Then I heard it again from a friend I had made at Penn, who, in advising me about the job I ended up taking, said, "Even when people say there's no hazing, there is. I'm sure you've heard that 'Nurses eat their young.'"

I have one personal experience with animals eating their young—the gerbils we had as pets when I was a kid. The two gerbils had babies. Then one day the gerbil babies began disappearing, until all four of them were gone. The only possible conclusion was that the gerbils had eaten the babies, and I remember as a young child feeling horrified when we realized what had happened. As a family we talked about why the gerbils would have done that and concluded that being trapped in a Habitrail had made them slightly crazy. We let those gerbils die a natural death, and then we never got any more.

The craziness that nurses experience has less to do with habitat—although a hospital can feel like a big Habitrail at times—than with situation. I read a lot of postcolonial theory as a graduate student in English, but you don't have to be familiar with *The Wretched of the Earth* by Frantz Fanon to understand that nurses see themselves as being on the bottom of the totem pole. Like people from low-status groups everywhere, some nurses take their frustrations out on other nurses rather than trying to improve their own position. It's not surprising that it happens, but it's especially poignant that people in a caring profession sometimes have such a hard time caring for one another. However, new nurses and inexperienced doctors are much easier targets than a system itself that does not value nurses' contributions.

On my old floor I tried to do an end run around the hostility and aggressiveness by keeping my mouth shut when I could. However, issues kept arising, like Irene needing morphine and breathing treatments, and in that particular situation, being submissive would have meant not giving her and her family the best care I could. For any patient at the end of life, when the lungs, or kidneys, or liver are shot to hell and death is imminent, there really is no place for arguments about paperwork, or for bullying, even if the bully is correct. When a beautiful woman gasps desperately for air in front of her loving family, when anyone is actively dying, we have to leave aside being correct and focus instead on doing good.

When I was trying to decide what to do about my horrible floor, my friend Julia, who's an employment lawyer, told me decisively: "Loyalty, schmoyalty—no one should yell at you at your job." The stakes are so high that maybe yelling seems necessary. Since the paperwork is involved with matters of life and death, it may come to be seen as important as life and death itself. However, on my new floor, where even more patients die with, at times, stunning regularity, no one yells, and the very important work we do as nurses always gets done.

TEN

Access

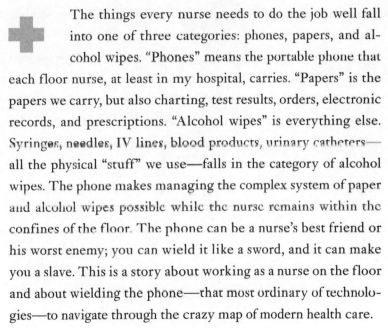

The things every nurse needs to do the job well fall into one of three categories: phones, papers, and alcohol wipes. "Phones" means the portable phone that each floor nurse, at least in my hospital, carries. "Papers" is the papers we carry, but also charting, test results, orders, electronic records, and prescriptions. "Alcohol wipes" is everything else. Syringes, needles, IV lines, blood products, urinary catheters— all the physical "stuff" we use—falls in the category of alcohol wipes. The phone makes managing the complex system of paper and alcohol wipes possible while the nurse remains within the confines of the floor. The phone can be a nurse's best friend or his worst enemy; you can wield it like a sword, and it can make you a slave. This is a story about working as a nurse on the floor and about wielding the phone—that most ordinary of technologies—to navigate through the crazy map of modern health care.

David was one of my patients that day. He had relapsed again, and the doctors wanted to give him more chemo. However, before David could begin treatment, he needed better IV access. That meant he needed a permanent, or indwelling, "line" that would make his veins constantly available to us. He had a peripheral IV, the kind anyone who has had a brief hospital stay or even same-day surgery will be familiar with. A peripheral line, usually called a heparin, or "hep" lock, is inserted by a nurse. She finds a good vein in the forearm or hand, sticks the patient with a needle, inserts a small plastic cannula into the vein, removes the needle, and finally attaches tubing to the cannula. Most intravenous medications can be given through this line, but the problem with hep locks is that they usually last for only two or three days. A patient getting chemo typically needs IV access for much longer, and many chemotherapeutic agents are so toxic that they cannot be infused peripherally.

I had four patients that day, but my most pressing task that morning was to get David down to the OR to have a triple lumen Hickman catheter—a specific kind of central line—placed in his subclavian vein. We were short staffed, and I had agreed to work a sixteen-hour shift. A long day, but in this case a good thing: it would take the entire sixteen hours, from 7:30 A.M. to 11:30 P.M., to get David his central line. Without that access he could not begin his new round of treatment, and cancer, unfortunately, does not wait.

David was what we would call a difficult patient. I'm not sure if he had an underlying psychiatric disorder or if his disease had turned him mean, but even though he spent his day reading the Bible and listening to Christian music, he would swear at staff, find fault with almost everything we did, and at times aggressively refuse care. He had relapsed two or three times already over as many years. Leukemia, or any cancer, can do that, and the doctors had decided what chemo to try this time. As the nurse, I had other concerns. David's recurrence of disease meant at the least another monthlong stay in the hospital, another struggle with nausea, hair loss, mouth sores, infection, or, if he was lucky, only some selection of the above, and another confrontation with death: would his disease kill him this time, and how severely would the treatment ravage his body?

The week before, David and I had had our own confrontation over IV access. He had been admitted to the hospital with an infection, and we were treating him with intravenous antibiotics. That night I was on evening shift from three to eleven, and David's hep lock had, as they do, stopped working. At my hospital we have an IV team made up of nurses who only insert IVs. I paged the IV team, and when they returned the page, I was told they would not come. The IV nurse said the patient had cursed at and insulted the team member who had come the last time. Not one of them on duty that night was willing to give David an IV.

Notoriety is never a good thing in a hospital, and this patient's behavior had made him notorious. Here's the problem, though—he had to have IV access. Without access, patients who need IV medications cannot get them. Just as important, if the patient crashes, and we have to call a condition, IV access is essential for getting fluids and drugs into him. It's dangerous to have a very sick patient on the floor without an IV.

I called the intern and told him about the IV team's refusal. He tried to put the IV in himself, but he had barely gotten the cannula in before the vein collapsed. Oncology patients often share this characteristic with IV drug users: weak, overused veins. The docs just don't get enough practice to be good at putting IVs in patients who are "hard sticks."

Now, though, the situation was untenable. The IV team refused to come, the intern couldn't put in the IV, David felt aggrieved about us poking him over and over again, and his IV antibiotics were getting more and more delayed. I was inexperienced enough that I didn't know what to do. I couldn't force the IV team to do their job, could I? I asked around, but no one had much of an idea how to work out this conflict. Finally, the intern suggested that we call the fellow, who came and said that if the IV team refused to come, we could put their refusal in the chart and say they would be responsible when the patient died of septic shock. I felt a surge of relief that finally someone with the experience—and the authority—to address the problem had come onto the scene. After offering that scary scenario, the fellow paged the IV team himself and told the nurse who answered the page that she had to come. His voice was calm and clear: "Hello, this is Dr. Suka. I'm the moonlighter, and you need to come put an IV in David Johnson." And that was it. The IV nurse came, David agreed to being stuck again, and the IV got placed without any drama at all.

A week went by, and David now needed the central line for his chemo. The IV team would not be required since David had to go to the OR, but sending an oncology patient to the operating room can be complicated. David's platelet count was low as a result of his disease. Without enough platelets, the patient's blood will not clot, making surgery that involves accessing large veins dangerous. So David needed a platelet transfusion, but he also needed some fresh frozen plasma because he was also low on clotting factors, the chemicals in the body that make it possible for blood to clot. I had given him a transfusion of red blood cells the day before for low hemoglobin, and he had a transfusion reaction that consisted of severe shaking, or rigoring, and chills. People are less likely to react to platelets and plasma than they are to red blood cells, but transfusion reactions can take the form of anaphylaxis, and since he'd had one the day before, he would need to be watched carefully while he got these blood products.

The previous day the intern, a tall, energetic brunette, had told me the plan for getting David his Hickman. I was to load David up with platelets right before the surgery, which was scheduled for 11:00 A.M. I had called the blood bank, explained the situation, and asked if getting the platelets would be a problem. I was told there would be no problem, although I wasn't sure how well the person I spoke with was listening. I had put the order for platelets into the computer, and I had explained the plan to the night shift nurse, who was supposed to start the platelets before I got there the following day. David's platelet count was hovering around 12,000, and they wanted him to be at 50,000 for the surgery. His body was chewing up platelets, so I needed to get them in fast to keep him at 50,000 just prior to the operation. It would be a race against the clock and his own body, and if his line was not placed, it meant another day without chemo, another day during which his disease could spread.

The shift started with a "Let the games begin" feeling. I had explained the plan to David. I had ascertained that he was in a bad mood, exacerbated by us not letting him eat prior to surgery. He still had just the one hep lock, and all of his products needed to go in through that line. When the first bag of platelets finished, I called the blood bank to ask where the second one was, and the plan the intern and I had created together crumbled before my eyes.

"A second bag wasn't ordered," they told me.

"But I put in the order myself."

"No, only one unit was ordered."

I pulled up the appropriate computer screen and saw that the order said three bags of platelets. "The order says three bags of platelets," I said.

"No, that means one bag."

"What?"

"A bag of platelets is a six-pack, so three bags means one," the person at the blood bank said. The intern was standing next to me listening to my end of the conversation. I looked at her and mouthed "Fuuuck," and she looked back at me wide-eyed. I hung up the phone and explained that, according to the blood bank, three means one, so only one bag of platelets was ready for David. She and I both paused for just a moment, sputtering to each other about how idiotic the system was. "So does four also mean one?" she asked me. But there wasn't time for indignation. It didn't matter that I had called the day before and the blood bank had told me that getting the platelets I needed would be no problem. I could waste time being angry and incredulous, or I could roll up my sleeves and get to work. The time was now, and we needed to solve this problem.

I called back the blood bank. "Can you send me the FFP so I can be getting that in while you get the platelets ready?" I asked. A more experienced nurse showed me that if you want to order more than one bag of platelets, you have to put them in as separate orders, exactly the opposite of how we order units of blood.

The blood bank sent the fresh frozen plasma up to the floor through our pneumatic tube system, and I slapped it up, or hung it, on David's IV pole, put the product number in the computer, and recorded David's vitals. Then the blood bank called me again. This time the blood bank physician was on the phone wondering if I really needed two additional bags of platelets in addition to the bag the patient had already received. The doctor was extremely kind but had a very thick accent, and I had a hard time understanding her. We had a long, winding conversation, during which she asked me what David's platelet count was and how much he usually "bumped," or increased, when he got platelets. Platelets are sometimes much harder to come by for transfusions than red blood cells; sometimes we even have platelet shortages in the hospital. Slowly, I figured out that the blood bank doctor wanted me to get David to 50,000, but not any higher, because another patient might have need of platelets, too, and 50,000 is high enough to make surgery safe.

Once I understood the competing needs, I agreed. "OK, two bags total; I'll draw a postcount and let you know," I said, meaning that I would have the lab come and take some blood to see what David's platelet count was after he received both bags of platelets.

I went to David's room to explain what was going on. Oncology patients who have had monthlong hospital stays and multiple relapses tend to be very savvy about their disease. David understood that he needed the platelets and the FFP. He was a tiny man in his fifties with scattered, spiky hairs growing on his head, giving him a punk rock appearance at odds with his close-mouthed manner and his Christian music. I think he actually liked having me as his nurse, but it didn't do much to make the day seem better for either of us. He was hungry and tired and did not like the idea of delays in going to the OR. "Hmph" was all he said when I told him the FFP would come first because we had to wait on the platelets. And then, "Well, I'm just really hungry here . . . it better not take too long," he told me.

Two of my other patients were also getting blood products, and the next hour became a swirl of checking ID numbers on platelets and blood, taking vital signs, and trying to get at least some of that information into the computer in real time. The OR had already called to tell me they were coming to get David. Surgery being at 11:00 A.M. really means anytime from 9:30 A.M. to 1:30 P.M., or even later, as I would discover. Surgeons, I thought. Their OR, their schedule. Or so it seemed to me. "He's not ready," I told the nurse who called, hearing the strain in my own voice. "He needs another bag of platelets." Sometimes if the patient isn't ready when the OR calls, the surgery gets canceled. No surgery, no line, no chemo—I made that phrase my mantra for the day.

The next time the OR didn't even call; they just sent up a tech with a stretcher. It was close to 11:00 A.M., and David's second bag of platelets had finished infusing. One of our roaming phlebotomists from the lab drew his blood for the postcount, while the tech waited with the stretcher. "I'll see you when you get back, David," I told him, "and you can eat then, too." He grunted, his lips held in a tight line. Then the OR tech took him down the hall, to the elevator, and away from me.

I turned to Maggie, the nurse working on the computer next to me, and speculated, "Do you think if I asked they would keep him down there longer?" I'd only been off orientation on my new floor for a month, and I didn't know this nurse well, but I liked the brightness in her eyes and her throaty smoker's laugh. She looked at me with surprise and said, "Oh, you are evil," then laughed her full laugh while I stood there feeling embarrassed. "Well, I was just kidding," I mumbled, too new to the floor and too self-conscious to get the joke. She looked at me again, still laughing. "Evil," she said.

Four hours later the OR nurse called me: David had not been

operated on. For four hours he sat, or rather lay, in the hallway. She said they were waiting on the postcount; that she called the lab numerous times, and the lab never called back; that David needed more FFP and they didn't have the nurses to hang it, blah, blah, blah. At this moment my impression of the whole day changed. I felt it come unstuck, rotate around in my line of vision, and reconnect, like a camera lens changing focus: a few clicks, and I had a new view of my job and of what I needed to do. David waiting four hours in the hallway, without food, most likely with minimal care: that was wrong. It happens, but it's wrong, and I felt like I needed to make it better.

I hung up my phone but kept it in my hand and pointed it for emphasis when I talked. "David has been down in the OR for four hours and didn't get his Hickman," I said aloud to everyone standing at the nurses' station. They looked at me in silence. Nurses are supposed to be the patient's advocate, and when something like this happens, it feels like a violation, like failure. No one had anything to say. I remembered my mantra—no surgery, no Hickman, no chemo—and I picked up the phone and started dialing.

I paged the intern, who was upset but couldn't do anything. Even though they're M.D.s, in terms of having power within the hospital hierarchy, the interns, and sometimes even the residents, often have less real authority than the nurses.

I paged the fellow. Since the fellow on night shift had smoothly orchestrated getting an IV for David the week before, I figured the hematology fellow could get David successfully back to the operating room. I had learned from the OR nurse that two different lab values were at issue: David's INR and his fibrinogen level. INR stands for international normalized ratio, and it matters because it tells whether a patient's blood will clot quickly enough that surgery can be safely performed. Fibrinogen is a protein needed for successful blood clotting. David's INR was at the level we had been told the surgeon wanted; concerns over his fibrinogen level were new, and I saw in that concern a possible opportunity to get David into surgery.

The OR nurse called me back. She wanted David out of her hallway and back on the floor to get his FFP. My concern was that if David came back to the floor, it would be very easy for them to put off his line placement until the next day, when the whole process of platelets, FFP, and not eating would have to start all over again. "His INR is where you told us you wanted it," I said to the OR nurse, beginning the negotiation with a strong lead.

"The surgeon wants his fibrinogen levels higher," she told me, ignoring the INR issue completely. "He wants the patient to get the FFP right before he comes down to the OR." The plan to give FFP immediately prior to taking the patient to the OR was new. "Can one of your nurses give the FFP down there?" I asked. "I'm just concerned that if he comes back up here to get the FFP, we won't be able to get him back to the OR to get the Hickman in him tonight. We've already given him platelets, he hasn't eaten—"

"No, we don't have the staff," she said, cutting me off.

I decided to pursue my initial tack: that they had changed the requirements midstream. "I thought the main issue was his INR," I said. The fellow, Edna, was standing across from me at the nurses' station, looking increasingly frustrated while she listened to my end of the conversation. A short woman with a pleasant but very determined manner, Edna reached out for my phone as I talked. "Hold on just a minute," I told the OR nurse. "I'm going to have you talk to the fellow." Giving Edna the phone was a risk. No nurse wants to suddenly be confronted by an M.D., especially on the phone, and though I did it to help David, I knew that it could backfire, making that nurse so angry at me that she just wouldn't find time for David in the OR schedule. However, by putting Edna on the line, I also let the nurse know that getting David his Hickman mattered to someone with more authority than I have, someone who potentially had the ear of an attending.

The nurse met my thrust with a deft parry of her own. She knew her job, and somehow she got the surgeon on the phone right away to address David's case personally with the fellow. Undeterred by the sudden appearance of the surgeon himself, Edna dove in, reeling off protocols related to INRs and fibrinogen values. I could tell it was not going to be a quick conversation, so I went to the computer to finish something else, keeping half an eye on her as she energetically argued David's case.

And then the conversation was over. David was coming back to the floor. He would get two bags of FFP and go back down to the OR—or at least that was the plan. My chastening came in the form of a phone call from the OR nurse. She wanted me to understand who makes decisions down there. "It's not up to the fellow," she told me. "The fellow can say whatever she wants, but it's the *surgeon's* decision; the *surgeon* decides whether it's safe for the patient to have surgery, and if the *surgeon* feels it's not safe, then the operation is not going to happen."

She had a lot to say to me, and in another situation, with less at stake, I might have spoken up for myself. I could have said that she left David in the hallway for four hours. I might at least have given into an adolescent impulse to hold the phone away from my ear and make faces for the entertainment of everyone at the nurses' station. However, no surgery, no Hickman, no chemo, so I listened to every word she said and tried to conciliate her. "You're right," I told her. "I think there were some misunderstandings," I said. "Of course," I agreed with her, and the truth is, she was right: the surgeon cannot perform surgery if the conditions seem unsafe.

Once she finished scolding me, it was over. Now we were working together. She told me the FFP was already ordered and on its way to the floor. David was on his way, too. She gave me her direct number and told me her name was Terri; I was to call as soon as the second bag of plasma finished. Terri and Theresa. Well, I had lost the first round of the OR battle, but maybe I would win the war. The people in the OR now knew that getting this Hickman placed mattered.

But the hardest part was still to come. David was at base an angry person, demanding, and anyone would have been upset after what he had been through. Plus, he still had not eaten, and being hungry makes people grumpy and aggressive. Our brains can only use glucose for energy, so hypoglycemia—low blood sugar—means a starved and irritable brain. The risk here was not that David would swear at me or insist that I leave the room. The risk is that he would need to assert his importance, let us know how aggrieved he felt, and that he would do that by refusing to go back to the OR that night.

Soon enough he was back on the floor. "How are you, David?" I asked him, trying to be empathic, but immediately realizing how banal my question sounded. "I'm hungry," he said, "I want to eat," then shut his mouth and held it in a grim line.

"OK," I said, pondering what to do, buying time. "Let me see about that," I told him. "Transport will get you settled, and I'll be right in." I quickly found Edna. "He's hungry," I said, "and he's angry, and he wants to eat." We keep patients "NPO"—not allowed to eat—before surgery because if they have food in their stomachs and they become nauseated from the anesthesia there is a very real risk of vomited food getting sucked into their lungs and causing pneumonia or respiratory distress. It's much more dangerous to operate on people who have eaten than on people who haven't, and surgeons insist that patients with scheduled procedures come to the OR with empty stomachs, even if they don't get operated on until eight o'clock at night.

"I'll talk to him," Edna said, and her face looked almost as grim as David's had when he returned to the floor.

"Do you want me to come, too?" I asked. She said nothing but crooked her arm toward me in an inclusive way. We began walking down the short bit of hallway to David's room when she confessed to me. "I just don't like him," she admitted. "I *really* don't like him." She shook her head, as if she were trying to clear out an unwelcome idea, and then she kept talking, almost to herself, looking straight ahead while walking. "But he needs treatment." I nodded. We had to convince David to hold out just a little longer so that the surgeon could place his line.

Once we got in the room, Edna started talking first, emphasizing the importance of the line and the necessity of starting treatment. David sat in his chair, having none of it. He was so angry he would not speak, and Edna's arguments seemed only to be increasing his resentment. He sat up even straighter in his chair and stared at her, giving no acknowledgment that she had spoken. This isn't working, I thought.

"David," I said, cutting in, "we understand that you're angry." He turned his unforgiving face to me, and I felt a sudden stab of anxiety over the need to pick my words carefully. "You've been through a lot today." I looked at Edna, and she nodded.

"Well, I was down there in the hallway, and they wouldn't let me eat." His voice was small and gravelly.

"Yes," I said, "and no one is saying that was OK or that you shouldn't be angry about it." I looked at Edna. "No, no one is saying you can't be angry," she affirmed.

"But here's the situation." I took a deep breath. "You need this line." David's eyes narrowed; he did not like what I was saying. I kept going, trying to put every bit of empathy I had into my voice, my face. I felt almost as if I were nothing but a voice, a dis-embodied voice in white scrubs. The rest of my argument came out in a rush. "If you can wait to eat, we just need to get some more FFP into you, and then you'll go down to the OR right away." I was offering him something I could not guarantee, but the tightness around his eyes eased up as I spoke, so I continued. "When you get back on the floor, we can get you some food." Again, what that food would be I could not say, since he would probably be off the floor until 11:00 P.M., and dietary would be closed. But he didn't ask, so I kept going. "You need this line, and if we can just get it done tonight, it will be finished. Otherwise, we have to start this whole process over tomorrow." I stopped, took a breath, and looked at him.

"Oh, all right," he said, sounding angry but also resigned. He looked down at the blanket covering his lap and picked at it. He would not look me in the eye.

"Great," I said, but I did not smile, and Edna did not smile, either. There was nothing here to be happy about. The situation had been a cock-up, but maybe we could fix it. "I'll be right back with the FFP," I told him. The clock was still ticking, and who knew what new wrinkle might intervene—in the OR or on the floor—if I dawdled.

Plasma can go in fast, and I set his pump to have it pour into him. His peripheral line held up, and within an hour or so the FFP was in. David and I talked very little as I came and went, checking his vitals, taking down the first bag of plasma, and putting up the second. What could I say? "I'm sorry?" "I'm embarrassed?" It's not like we had endangered his life, but we also had not taken care of him the way we should have. Maggie, the nurse who had joked that I was evil, came to David's room to check the plasma before I hung it. I complained to her, "The OR nurse said, 'He's been sitting in the hallway for four hours.'" And then in a rush of anger I blurted out, "Not on my floor!" Maggie looked at me. "Well, that's right," she said. "They're all nurses down there."

As soon as the FFP had finished, I called Terri and told her David was ready to come down. I put him in for transport. I announced, to no one in particular, that if they didn't put that Hickman in, I was going to come down to the OR and put it in myself. This earned me several cautious looks, as if my dedication was admirable even if in my zeal I seemed a bit deranged.

I had hoped that David going to the OR would come off smoothly, but, of course, there were a couple more glitches to get through before he left the floor. First, Terri called me back and told me David would have to wait; an emergency had just come into the OR. I wanted to say, "You are fucking kidding me!" but instead I nodded, spoke calmly. She promised to call me back in five minutes with an update.

Meanwhile, transport had arrived, and the transporter was getting David onto the stretcher ready to go down. I asked him if he could wait the five minutes until Terri called back. Usually it takes five minutes just to get the paperwork printed, the forms filled out appropriately, the patient settled, the IV pump ready to roll along. However, this particular transporter was slightly mentally retarded, and he was due to go off work in four minutes, at 8:00 P.M. When I asked him to wait the five minutes until Terri called back, he said, sounding panicked, "I have to leave at eight. My shift ends at eight." At this point, if I had had the time, I probably would have gone into a room and screamed. I would have taken my phone and thrown it, hard, grinning crazily as it broke into a hundred pieces. I would have smashed the transporter up against the wall, Dirty Harry style, and told him that tonight he was staying long enough to get my patient to the OR.

But none of that ended up being necessary. Terri called me back. The emergency was under control, and they were ready for David. The transporter's boss gave him permission to stay a few minutes past eight o'clock, and at two minutes to eight he wheeled David away.

David finally came back to the floor at 11:30. My long shift was over, and I was spent. The nurse taking over for me settled him back into his room and found him a sandwich in the kitchen on our floor. He was groggy from being sedated and not ready to eat, but when he came out of his stupor, he would be hungry. I didn't talk to him, and in fact I barely saw him, but a glance as the transporter wheeled him past was enough to let me know: at last, he had his Hickman.

There's a great Rudyard Kipling story from *The Jungle Book* called "Rikki-Tikki-Tavi," and the story came to mind as I thought over my day. Rikki-Tikki-Tavi is a mongoose that gets washed out of his nest during a flood and ends up living with and protecting a family of British colonials in India. Father, mother, and son live in a house that a cobra couple also has their eye on, and the cobras decide to kill the humans so that when their eggs hatch, their babies can roam freely. Rikki is a young mongoose, but he manages to kill both cobras, saving the British family. The last lines of the story applaud Rikki's success and describe his lifelong vigilance: "Rikki-Tikki had a right to be proud of himself. And he kept the garden as a mongoose should keep it, with tooth and jump and spring and bite, till never a cobra dared show its head inside the walls."

Traditional views of nurses portray us as ministering angels, selfless guardians in floral scrubs. We can be selfless, and patients have called me an angel, but in a situation like David's, an angel has much less to offer patients than a mongoose. Rikki-Tikki attacks snakes with a rocking motion "so perfectly balanced that he could fly off in any direction he wanted." When the cobras antagonize the birds in the garden, "Rikki-Tikki felt his eyes getting hot and angry, and he sat back on his tail and hind legs like a little kangaroo and chattered with rage."

I keep my patients safe with my own version of tooth and jump and spring and bite: I wield phone and voice and brains and grit. I am pulled in multiple directions every day, and though I cannot claim perfect balance like Rikki-Tikki, I keep my feet pointed toward the greatest need. I do not have time to chatter with rage, and no one else would have time to listen, but I have felt my eyes get hot and angry when my patients are not treated as I would like, and I will fight to make things better.

I was off work for two days after my sixteen-hour shift, and when I came back, David was not my patient. However, I saw him standing at the nurses' station asking for stamps. "Hey, David," I called out, passing by on my way to do something else, "how's that Hickman?"

He turned to me and smiled an amazing smile, a smile so full of light and happiness and even joy, that for a moment I wasn't sure I had the right patient. "Oh, it's great," he said, still smiling. "I'm so glad," I told him, smiling myself. And then we nodded to each other and moved on.

ELEVEN

Poison

No one wants to think that out of the blue they could be diagnosed with cancer, but that's pretty much how it always happens with leukemia. A patient will come in and say she had been feeling really tired for a while, or he noticed that he bruised easily, or she had a cold that just wouldn't go away. It could be any of us at any time. First, the doctor tells you that you're anemic, and she treats that. Then, when the exhaustion continues, the doctor orders more blood work and does a biopsy of your bone marrow. The call might come when you're grocery shopping—a danger, perhaps, of having a cell phone— "This is Dr. Jones, and you have leukemia. You need to go to the emergency department of —— Hospital to be admitted. They know you're coming. And you need to go right now."

Do not pass Go; do not collect $200. I can only imagine what it must feel like to get that call. Standing in the aisle with the canned tomatoes, peaches, and beans, the world must turn upside down. I envision cans spinning around in a kaleidoscope of color as the life you thought you knew becomes suddenly alien and unsafe. "You have leukemia," and it will take a long time before the world comes to rights again, because the cure for this kind of cancer is almost, but not quite, as bad as the disease. Patients know they will be treated with chemotherapy, but the reality that chemo is, literally, poison won't sink in until they're already in the middle of their treatment.

It could be any of us who has to endure this disease and its harrowing cure, but I'll tell the story here of my patient Bill. Bill had actually been fairly symptomatic. A blue-collar guy who worked in a warehouse in rural Pennsylvania, he had been so exhausted for the few weeks before his diagnosis that learning he had leukemia was almost a relief: at least then he knew something was definitely wrong with him. Bill was in his fifties, his children were adults no longer living at home, and his wife was sweetly attentive. He had a gentle manner and a direct but polite way of speaking that reminded me of my former physiology professor at Rutgers, another rural Pennsylvanian. The resemblance mattered because it made me warm up to Bill right away. This could be Henry, I thought, and remembering the professor who had made human physiology so interesting, and imagining him in Bill's position, I felt protective.

Bill's wife was another matter. She spoke slowly and deliberately, but her anxiety over him made it difficult for her to absorb what we were telling her. Her worry, in a sense, made her stupid, and she would ask the same question over and over again, each time in the same slow cadence with the same deliberate articulations. This was maddening, especially when it happened over the phone, but we all knew that she was adapting to a lot, too. Her husband had been feeling severely fatigued, true, but it had never crossed her mind that he might have leukemia. Why would it? So now he was in the hospital, an hour's drive from their home if the traffic was good, and she was alone dealing with the work around the house, paying the bills, staying overnight at the hospital (we have twenty-four/seven visiting hours), going back and forth to transport dirty clothes and bring in food Bill might like, and encountering a rotating staff of nurses and doctors who did our best to keep Bill's story straight but were in no way perfect. She was annoying, but she was coping, and she was helping him cope. They were high school sweethearts and had been married for over thirty years. To have that bond suddenly taken, broken apart, then tied up in a knot and tethered to a hospital bed—any of us would do well to be occasionally annoying and no worse.

When patients like Bill arrive at the hospital, their moniker becomes "new diagnosis AML," for acute myelogenous leukemia. AML is the most common blood cancer among adult patients, and it is treatable with chemotherapy. However, for most people, the treatment is physically demanding, and regardless of its success or failure, treatment means an initial hospital stay of six weeks with no promise of a cure when that six weeks is up.

The treatment for AML is standard and comes in two parts: induction and consolidation. During induction, patients receive large doses of chemotherapy designed to bring on a remission, or a reduction of the disease to levels that cannot be detected. Consolidation involves lower doses of chemo given at set intervals over several months. The idea behind consolidation is to achieve a "cure," the holy grail of cancer treatment. Induction for AML consists of two drugs, cytarabine, which we call ara-C, and an anthracycline type of chemo, usually either idarubicin or daunorubicin. Because everything in medicine gets shortened, we call this regimen "7 & 3," because the cytarabine is given continuously over seven days, the idarubicin for only three days in a much smaller dose that comes as an "IV push," which means more like a shot. Another shorthand notation is "induction with ida and ara." Once a nurse has been on the floor for a while, these telegraphic communications start to make sense, but in the beginning it sounded like gibberish, and scary gibberish because the drugs are so toxic.

The first time I gave chemo, another nurse supervised me closely, walking me through it step by step. First, I got the drug out of the special bin for chemotherapy. Pharmacists in the oncology satellite pharmacy prepare all the chemotherapy, then leave it for the nurses to administer. All chemo comes wrapped in an outer bag, really just a large Ziploc bag, taped with a caution strip: "Warning: biohazard drug, handle appropriately." The drugs have their own aura, which waxes and wanes for me but never completely goes away. I doubt that I will ever feel blasé about giving chemo, and the process we go through to administer the drugs helps ensure that aura will remain. Two nurses "check" the chemo, always. One holds the bag of drug and reads what it says, while the other compares that information against the order in the patient's chart: "Patient John Smith, mitoxantrone, 13 mg, given as a fifteen-minute infusion, day 3 of 5," or "Jane Johnson, cyclophosphamide, 250 mg, continuous infusion, day 2 of 6." The dates on which each drug will be administered are also part of the order, and after both nurses double-check the drug against the paper order, they initial the order on the back under the appropriate drug and date. Some "drug regimens" have developed over time through long processes of trial and error in cancer research; other regimens, like "7 & 3," have remained essentially unchanged since they were first introduced, in this case in the 1970s.

There's a lot to learn, and the oncology and chemo classes I attended at work only touched the tip of the iceberg. Like all drugs, chemo drugs have at least two names, and people use them pretty much interchangeably. Cytarabine is also ara-C, vincristine VP-16; cyclophosphamide is cytoxan, and imatinib mesylate is Gleevec. The regimens have names, too, and those can seem random as well: CHOP, a standard therapy for non-Hodgkin's lymphoma, is **C**yclophosphamide, adriamycin (generic name **H**ydroxydoxirubicin), vincristine (brand name **O**ncovin), and **P**rednisone. Sometimes CHOP has an *R* added to it for Rituxan, and if Rituxan is ultimately not part of the regimen used, we end up calling it "R-CHOP without the R." None of it makes intuitive sense, and the attendings have spent years learning these regimens, knowing when they're indicated, what the risks are, what adjustments need to be made for weight, age, past medical history, and possibly cytogenetics. I did my oncology and chemo training when I was on light duty, progressing slowly week by week from crutches and ice packs to weight-bearing activity, while my bewilderment over these treatment choices only increased. I returned to the floor able-bodied and chemo certified, regaining my sea legs as a nurse with the additional responsibility of learning this important and scary task.

So I found myself a couple of weeks after I got back on the floor getting tutored by another nurse in the fine points of giving chemo. Together we checked the order and then had another nurse check it because I wasn't yet considered qualified. I put on the special blue full-body chemo gown and the required two sets of blue chemo gloves, the first set under the cuff of the chemo gown, the second set over the cuff. Christy, the nurse training me, said, "We're supposed to wear two sets of gloves, but I can't feel anything with two sets of gloves on." Still, she dutifully pulled on a pair of gloves over the pair she already had on her hands. She told me to grab a couple of syringes with saline and bring the bag of chemo into the patient's room.

I introduced myself to the patient, whom I had never met, and explained that Christy was teaching me how to administer chemotherapy. The patient was an older man in his sixties, and my newness didn't faze him at all. He was an old hand at chemo and could probably have told me more about the drugs he was getting than I knew at that time. Certainly, he knew more about his diagnosis and prognosis than I did—I was only giving him the drugs that just might save his life. Christy and I did everything by the book: checked the patient's name on the chemo against the patient's wristband, looked at his central line and made sure I had an open lumen for giving the drug, then used one of my saline syringes to make sure I saw blood, or "got a blood return," from that lumen, our procedure for making sure the line was still in a vein. After all that I "hung" the chemo: took the drug out of its outer Ziploc bag, put it on the patient's IV pole, strung it through the IV pump, connected the end of the IV tubing to the patient's line, programmed the pump for the rate of administration specified on the bag, turned it on, and let it go.

The thing I remember most is that by the end I was sweating. I had been nervous, and the blue gowns we wear for protection are plastic and hot. The patient thanked me and said I had done a good job, a sure sign he had spent way too much time on the floor since he knew our routines and rules about chemo as well as many of the nurses and better than I did. Still, I had done it,

I had given my first chemo; I had been initiated into the potentially life-saving part of oncology nursing, and it gave me a rush I hadn't anticipated. This is power, I thought, and the realization scared me. As a nursing student, in anatomy class, I held a human heart in my hands. The heart is smaller than I thought it would be, and it looks so simple: four chambers, an opening for the aorta, the pulmonary arteries and veins, the subclavians. This elegant structure that never stops beating gives life to all humans, and most animals, on earth. I remembered a feeling of power as I held the heart in my hands, but an even stronger feeling of awe. Chemo is both less elegant and less universally effective than the human heart, but in addition to a feeling of power it also invokes a sense of awe, not the least of which is the trust patients show nurses when they allow us to inject these poisons into their veins. Giving chemo is a huge responsibility, and I try not to ever be so rushed that I forget all that I'm responsible for.

These thoughts, feelings, and memories came back to me as I met Bill for the first time. He had already had his central line placed, and his MUGA (multiple gated acquisition) scan had come back telling us that his heart could tolerate the standard dose of ida and ara for induction chemo. The MUGA scan measures cardiac function, especially how well the left ventricle works, because idarubicin can be toxic to the heart. Patients getting ida always need a baseline MUGA to show how well their heart is working before they begin treatment.

Bill had been through all this in a blur of trips on and off the floor, and now, looking stunned, resigned, and hopeful, he was ready to begin chemo. Giving the doses themselves wasn't any different from any other time I gave chemo. Cytarabine hangs as a twenty-four-hour infusion, with a new bag going up once a day for seven days, and idarubicin is given as an IV push. The ida is orange, and the nurse administers it by injecting it into running saline, alternating drug and saline to make sure that the color in the line never gets too dark, ensuring that the patient gets the drug diluted. Going in, idarubicin looks like the orange soda I drank too much of when I was a kid, but coming out of the body it can turn patients' urine red as blood, an effect they will find disturbing if they have not been forewarned. Pushing ida means sitting in the room the whole time it goes in. The nurse who taught me to do this said to pull up a chair because it would take a while.

When I gave Bill his treatment, most of these details were fuzzy in my mind. I knew what drugs had been ordered for when, and I knew the rudiments of giving them. IV push chemo still made me very nervous, which turned my time in the room sitting in the chair into an awkward silence. Or maybe the patient felt awkward—I only felt very focused on what I was doing, and I'm sorry to say that Bill, and what might have been going through his head, was not much on my mind. I put a special chemo pad under his line where I injected the idarubicin and watched the color in the line as I pushed in the drug, let the saline run, and every so often pulled back on the syringe to make sure of that all-important blood return.

Patients with a new diagnosis approach chemo in usually one of two ways: with nervous apprehension or a tense eagerness. One day I had prepared to give chemo when the patient I was administering it to said, "What does it mean that you're all dressed up in that special gown and gloves, and you're shooting that stuff right into my veins?" "Well," I said, "with us they worry about cumulative effects," which is true, but only half true. These drugs are dangerous biohazards. On my floor, pregnant nurses and nurses who are breast-feeding do not administer chemo. Our gowns and gloves, and the empty containers of drugs once they've been administered, get thrown away in special yellow trash cans marked "For chemotherapeutic waste only." If chemo gets "spilled" (an event I've never seen), special procedures exist for cleaning up the mess.

For every patient who is apprehensive, another will be in a hurry for the chemo to start. "What's the holdup?" some patients will ask, and "Why the delay?" which evokes a long explanation of how the dose has to be tapered to their weight and height, checked and double-checked by nursing and by the pharmacists and doctors, then mixed, dropped off, and double-checked again. Bill hovered just this side of anticipation. He was a patient man, and I got the feeling that not much bothered him, even the waiting, which was lucky for me, since my ability to explain what I was doing was at that time sorely limited. On the day of his first chemo treatment, I checked his line for a blood return and gave him his drugs. Then I had the nurse who had initially checked the drugs with me come into the room and make sure I had programmed the correct rate for cytarabine into the IV pump. After that, she and I checked off the drugs on the computer together, since chemo needs to be validated by two nurses, and the witnessing nurse has to type in a password unique to approving chemo.

All of these procedures have now become more or less routine to me, but at the time I treated Bill, I had to think through every step to make sure I remembered it. There were other things to remember, too. I gave Bill patient education handouts on the drugs he was getting. I checked how long his urine would be biohazardous and posted a notice with that date on the door of his room and on his bathroom door. Thirty minutes before the chemo, I premedicated him with Zofran—an antinausea med—and dexamethasone, because steroids and standard antiemetic drugs have a synergistic effect that leads to even less nausea for most patients. A lot of things have to be done to administer chemo in a way that is safe for patients and staff, and on busy days I often felt anxious about forgetting something.

However, things went smoothly for Bill, who got his drugs and his patient education handouts on time and finished up his seventh day of cytarabine. Then the waiting began. The problem with treating leukemia is that eradicating the disease involves making all the problems already caused by the disease even worse. Patients with leukemia often have suppressed immune systems, are anemic, and have greatly reduced numbers of platelets; hence the symptoms include colds that never seem to go away, fatigue and sometimes shortness of breath, and easy bruising. The medical word to describe all these problems is *pancytopenia,* or a severe drop in blood counts. Leukemia causes pancytopenia because the bone marrow is producing leukemic cells—immature white blood cells called "blasts"—instead of mature white cells, red cells, and platelets. Killing off the leukemic cells typically means that *all* of a patient's normal blood cells will drop in number. The lowest point of this drop, the nadir, comes approximately ten days to two weeks after the end of chemo. Patients' hemoglobin and platelets will drop low enough that they need transfusions to stay alive. Their white blood cell count will reach zero, meaning that they have no effective immune system and are extremely vulnerable to infection. At this point in the treatment arc, patients have become savvy enough that they begin each day asking, "How are my counts?" and often request printed copies of their blood work. At this point also, the most difficult part of treatment begins.

Bill had a very tough time. Some patients getting chemo develop terrible mouth sores simply as a result of the drop in their counts. We call it "mucositis," but that benign-sounding name does not capture how intensely painful those sores are for patients. Bill's mouth became so agonizing to him that he stopped eating. He had to be on a permanent morphine drip, with additional doses he could administer himself, just to keep the pain under control. We recommend that patients rinse with saline after meals and before bedtime to reduce the chances of developing mouth sores. Bill had done that religiously, but the sores, as they do for some patients, came anyway. He started to lose weight and stopped taking his walks around the floor. We gave him antiseptic mouth rinses and antifungal swishes, but nothing helped with the pain, and our swish-and-swallow medications just made him feel like he wanted to vomit.

Still, he remained his usual calm self. Whenever I went into his room, he would always ask how my kids were. He reassured me that good parenting is a matter of good instincts, and he believed I had those. The stock market was in free fall, and he had to sit helpless in his hospital bed as his 401(k) dwindled away and he realized how much longer he would have to work once he left the hospital just to have enough to retire. We talked about his anger that he had been a responsible home buyer, but now his retirement would be compromised by people who had not been nearly as prudent.

Bill had been in Vietnam, mostly in the rear echelon servicing helicopters, although he had been shot at a few times and described the experience as "very scary." He wondered if his leukemia could be the result of Agent Orange exposure, and I hooked him up with a social worker who could investigate that question with him. He watched CNN a lot, and at one point I was in his room when bad news about the Iraq War came on. I asked Bill if he thought President George W. Bush regretted all the destruction he had unleashed in the Middle East and the many dead American soldiers. "No," he said, waving his hand at me dismissively. "He's the kind of guy who never feels bad about anything." It was interesting to me that postmortems on the Bush presidency came to a similar conclusion.

His wife came and stayed over, and I dealt with her repetitive questioning as best I could. I guess it worked because at one point she asked me, "Are you always this nice?" to which I mumbled an embarrassed and tongue-tied "Well, people do say that I'm nice," not knowing how else to answer.

Bill's mouth was still bothering him, but he had started eating again when the storm really hit. He developed a neutropenic fever (neutropenia means a patient's white blood cell count is low), which is pretty much expected in leukemic patients receiving chemo. Blood cultures revealed a serious infection, and the doctors increased the variety and dose of his IV antibiotics. Not only was he getting antibiotics in bulk, he was also getting vancomycin through each of the three lumens of his central line at the same time, a sometimes controversial treatment meant to preserve the line in cases of infection. He lost more weight, slept most of the time, and talked little.

Then one day I came in and saw him making a slow walk around the floor. "Hey, Bill, you couldn't have done that yesterday," I said. "Yesterday I didn't feel up to it," he answered, and I explained, "That's the point—yesterday you couldn't have done this, but today you can. That means you're feeling better." He stopped for a minute and nodded, absorbing what I had said, then kept on walking. Neither good intentions nor disease would deter him for long if he could help it.

His counts were going up, the fevers had stopped, and the attending, after Bill had been in the hospital for six weeks, said that maybe Bill could go home that Friday. "If your counts keep going up, and you don't spike any more temps," the doctor told him. Then the diarrhea started, and all bets for discharge were off. Bill started having profuse malodorous stools and tested positive for *Clostridium difficile*, which typically strikes after patients have had their normal intestinal flora killed off by antibiotics. On my floor we rotate around depending on the staffing demands and who has been where on the floor for previous shifts. I now rotated away from Bill, and though I regretted not being there for him, I was also relieved. His mood was spiraling downward. I understood why, but I found it hard to take mostly because I found it infectious. In my heart I thought he would be OK in the end, but his despair at one setback after another and his disappointment at not going home when the doctor said he might were influencing me, too, and I found myself unable to comfort him in the way he needed comforting.

So I watched from a distance and got reports from the other nurses and from his wife, whom I would run into in the patient kitchen or on her way to the hospital cafeteria. This was one time when I understood why she perseverated. "If only the doctor hadn't said he might be going home," she would repeat over and over again to me. Bill had gotten set on being discharged, and when the diarrhea started, it really dashed his hopes. His mood sank, and his bowels locked up. The doctors were afraid his intestine might perforate because once we cured the *C. difficile* infection, he stopped having peristalsis at all—in an unusual variant of *C. difficile,* Bill had developed an ileus, meaning that part of his small intestine just was not working. I was very busy with my own patients those days and heard only vaguely what was going on with Bill, but the details were not good: really sick; might need a colostomy; getting nutrition through his IV; still losing weight; very, very weak. Someone even raised the idea of a "stool transplant," where feces from one of Bill's family members would be inserted into his rectum in an attempt to recolonize his GI tract and get his gut working again. I could tell I was withdrawing, not wanting to hear the bad news, smiling empathically when I saw his wife in the hall, but not doing much more than that.

Then one night I showed up for a three-to-eleven shift, and Bill was my patient. By this point his ileus had resolved, and he was eating again. The worst was over, but his mood seemed to have turned permanently sour. He grumbled about medications and took any suggestion, even at times conversation, as an affront. His wife was finding it difficult to be in the hospital with him. I understood it. He had been living his life when we took him, hijacked his ordinary existence, and in the name of saving him made him so weak and so vulnerable that really we almost killed him. He'd found the diarrhea debilitating and embarrassing. For two and a half months he'd been hooked up almost permanently to an IV, and having to take that pump with him every time he wanted to take a walk, or even go to the bathroom, made him feel like an invalid and a slave. His room was

big for a prison cell but as small as a Manhattan studio apart-
ment, and he'd been stuck there for too long. He rarely got outside and missed breathing fresh air. He'd lost weight, some fat,
but muscle, too, and when he looked in the mirror, he saw only
a ghost of his former self. A bone marrow biopsy had shown that
his cancer was in remission, that he was "clear" of blasts, but so
far he had not been healthy enough to absorb that good news,
and now all he wanted, even if he had been cancerous, was to
leave the hospital and go home.

He was close. He'd been retaining fluids, which gave his
medical team some concerns about his heart and his kidneys,
but the plan was to get him off IV fluids and see if the edema
resolved. So he was off the IV pump, and he no longer needed
constant doses of intravenous antibiotics, just vancomycin by
mouth. He was eating, and his appetite was growing, especially
when Joanna, one of our housekeepers, brought him homemade
lasagna.

It seemed that he was pretty much done with all of us, that he
tolerated us as a necessary evil until he could leave. Imagine, two
and a half months of having your vital signs—blood pressure,
heart rate, oxygen level, respiratory rate, and temperature—
taken every four hours awake or asleep; of never having any
privacy; of trying to pay bills and help out your wife when the
furnace breaks down; of being in touch with the outside world
only through your television and your cell phone; of having to
ask for everything you need: ice, clean bed linens, fresh water, a
Tylenol. For Bill to take a shower, a nurse would have to disconnect him from his IV and then tape a plastic cover over the IV
site. Once he got out of the shower, the whole process had to be
done in reverse. His meals came up when they were ready, not

necessarily when he wanted them, and he had been in the hospital so long that everything tasted the same.

However, we had cured him. He would need consolidation chemo over the coming months to cement his remission, to really be cured, and he could relapse, and he might need a stem cell transplant down the road, but for now he was going to be OK. We had saved him, but in a way I felt we had failed him, because the process of saving him had made him so miserable, had made him hate the life he was living.

The only thing I like about doing the three-to-eleven shift is that we get to tuck the patients in, and they're often asleep, or close to it, when night shift takes over. It was near the end of my shift, and I got Bill settled in bed for the evening. He was off the IV pump, had eaten a real dinner, and was afebrile and free of pain. He had adamantly refused antidepressants, antianxiety meds, and even sleep aids, but he had walked the halls today, and I hoped that exercise would be enough to give him a good night's sleep. Sound sleep is often hard to come by in the hospital, and his two months' accumulation of sleep deprivation had not helped his mood or hastened his recovery.

I was just leaving, saying my good nights, when Bill spoke. "Thanks, Treese," he said, because he'd given me his own nickname. "I needed a dose of you tonight." It was one of the nicest things anyone has ever said to me, certainly one of the nicest things a patient has said, and it restored my faith in the work we do in oncology. He wasn't angry at me or at us, but at the situation, at his disease, at his imprisonment in this house of healing, at having to lose his dignity and his sense of self in order to save his life.

Chemotherapy is a Faustian bargain, a deal with the devil. The luckiest patients, we say, are the ones who are bored, who aren't dealing with intolerable mouth pain, or nausea and vomiting, or odd rashes, breathing problems, infections, cardiac troubles, nerve pain: the list of side effects is almost endless, and all of the side effects are bad. A few patients sail through, and we hold them up as examples, to give hope to the rest, to say, "Hey, it may not be so tough." But I would hold Bill up as an example, too. He went through hell, but at least for the moment we cured his cancer and saved his life.

His discharge had been pending, and the next time I came to work, he was gone. Just like that he went home, out of our care and out of our treatment range forever, because I never saw Bill again. Probably he got consolidation on our sister unit, where I used to work. Or maybe he somehow got it inpatient nearer his home. I don't ask, because to me it's important to remember the good but horrible work we did for him when he was first diagnosed. Maybe he'll be one of those people who come back after two years, walk onto the floor, and tell us with happy disbelief that it's been two years and they're cancer free. I met one of those patients when I first started working as a nurse. I didn't know him, but the other nurses looked on him as a small miracle. I hope that one day Bill comes back to be our miracle: Bill or someone else. It's good to know that all this suffering can actually keep our patients alive.

Epilogue

 In Barbara Kafka's classic cookbook *Roasting: A Simple Art,* she advises:

When you're hungry, roast.
When you're in a rush, roast.
When you're in doubt, roast.
When you're entertaining, roast.

In my own mind I extended each of these lines to read "roast a chicken," because I liked the idea that roasting a chicken could be the solution to all of life's problems. When you're unhappy at work, roast a chicken; when you just wrecked your car, roast a chicken; when you're mad at your husband, roast a chicken. It takes ninety minutes to roast most chickens, and I thought that after stuffing a lemon and some thyme inside the bird, basting the outside with butter, and scattering chopped onions around it to brown, whatever troubles were pressing would seem less so. When the succulent aroma of the roasting chicken began to fill the air, a feeling of ease would return to the day. Unless you're a vegetarian, nothing says home and safety quite like roast chicken.

However, there are troubles in life that no number of roast chickens can lessen, and the foremost of those is death. Death. It casts a long shadow in this book, and in these stories. Even when death is not present it hovers just around the corner, unbidden and unwanted, but waiting nonetheless.

I would like to end this book with a simple list of names, but restrictions on patient confidentiality prevent me from doing that. I can continue to tell people's stories, though, and explain what I have learned from them. The list of the dead is long, and each life was unique. I cannot tell all their stories, but I can share a few.

There was Barney, who had been on the floor for so many weeks he began calling himself "The Mayor." In his midforties and basically healthy, he loved to walk in the mornings and would be up and at it before we had even begun morning shift at 7:30 A.M. He was a big kidder, and when his hair began falling out he wore a wig with dreadlocks on his walks. Other times he put a big lipstick kiss on the outside of the yellow paper isolation mask he had to wear. I would call out, "Who's been kissing you, Barney?" and he would yell back as he kept walking, "The plumber—he can't keep his hands off me." His wife was, truly, as cute as a button and plied the nurses with Halloween candy, homemade banana bread, and her own special sausage pierogies. Now he's dead.

Then there was Jane, a nervous thirty-something with a history of schizophrenia. Time and time again she was hospitalized with one or another infection. She once had a hacking cough loud enough to hear over the entire floor. Nervous and fidgety, she often seemed ill at ease in the hospital, but during her rare walks with her parents—who were kind and attentive—she would smile if I called out her name. "Hey, Jane, it's good to see you up. You look great." Her face would brighten, and as she smiled, her restlessness would vanish into the happy crinkling of her eyes. As a result of her disease and her many treatments, she couldn't live without blood transfusions, and after one infection too many, she died.

There was the redhead in her late fifties who was feeling pretty good and really wanted to wash her hair but couldn't because of a bandage on a surgical site. With jokes about swearing her to secrecy, I got her a "shampoo in a bag," a toiletry so rare (I guess) that it is kept under lock and key separate from the other soaps, shampoos, and lotions we have. The patient was supposed to crinkle the bag, which looked like a shower cap, put it on, stuff all her hair inside, and massage the "waterless soap" into her scalp. Afterward she said it wasn't as good as a real shower, but it sure was better than not having her hair washed at all. Two months or so later she was my patient and on hospice, her face so hollowed out that I could barely recognize the determined woman her entire family called "Red." She died at noon that day.

Then, of course, there are the old people, the wasted bodies who've simply reached the end of the line. Those deaths sit a little easier on all of us, since they don't raise concerns about lost potential or what might have been. Those deaths often come as a relief to everyone. But there are other deaths I have not stopped mourning, that have changed who I am and how I look at the world.

Two of these deaths occurred at roughly the same time on the floor. In some ways, except for their different genders, the patients were interchangeable. They were both middle-aged with adult children and grandchildren. They were both constantly visited at the hospital, by relatives and by friends, and without exception the people who visited them were pleasant and cordial and seemed to really care.

The man, Tom, had come in with leukemia, and his first round of chemotherapy had not killed off his cancer, which we knew because his bone marrow had been sampled and was not "clear," or free from disease. Following the advice of his doctors, he decided to go through another round of chemo. His wife was often at his bedside, and one day when he was asleep, she and I began talking about this. She began to cry. "I'm just so scared," she said. "What if he doesn't make it this time?"

I did what nurses so rarely have time to do: I sat down, held her hand, and listened. Eventually she shooed me away with a wave of her hands, saying, "You girls have a lot to do," but before I left the room, I looked back at her. "Are you all right?" I asked. "Yes," she told me, but how all right can anyone be while worried about her husband's death?

I had taken care of Tom a lot during his first round of chemo, but was assigned to other patients for this next round. I hadn't seen him after the second course of chemo started, but one day the nurse who had him called me into his room because she needed help. I walked into the room and did not know this Tom. He was drooling uncontrollably and had little control over his arms or legs. He did not know who he was or where he was. I could not believe the transformation. Turns out we had obliterated his immune system, which in some ways is the goal of chemotherapy, and he now had three infections in his blood. Later that day he went to the ICU, and a few days after that he was dead.

His female counterpart, Anne, had a very similar story, except that I met her right at the start of her struggle with cancer. I admitted her to the floor, and she really was the picture of health. She had lots of energy, no complaints, a hugely positive attitude and was accompanied by her husband, who seemed to get a big kick out of being with her. Every day the two of them took long walks around the floor, and she continued to look healthy and energetic. As her treatment progressed, I again got assigned to other patients, but next thing I heard was the chemo had made no difference in the progression of her disease, and she was considered "end stage," which is how we now describe patients who were formerly called "terminal." She also had pneumonia and a fever that we couldn't make a dent in despite the arsenal of antibiotics we were using to treat her.

The phrase "spiraled down" is overused, but in Anne's case it applied. Over the next few days her breathing got worse until she could not breathe on her own without a high dose of oxygen. Her fevers never subsided, and she also was sent to the ICU. A few days later a patient I didn't even know Anne had been friendly with stopped me in the hall and said, "Did you know that Anne died today?" I refused to believe it. "When I admitted her she was fine," I kept saying to no one in particular. "I mean, I know she had cancer, but she looked totally fine." An older, more experienced nurse finally heard me. "That's what's hard," she said.

What I wanted to say was "No." I did not want to think of Tom's wife or Anne's husband grieving for their lost life partner. I did not want to think about the vacuum left in the lives of their children, grandchildren, and friends. Tom had a daughter who had just gotten married, Anne a son and daughter-in-law who were expecting twins. These oh-so-important life events would be forever colored with grief. Start to finish, for both of them, it had been roughly one month, maybe two, from admission to death.

Soon after the deaths of these two patients, my husband's paternal grandmother died. This odd woman became his father's stepmother after the death of his biological mother when he was very young. Mollie's death was not a sad passing. If she had grown up in this day and age, she would probably have been diagnosed and medicated. Her paranoia took the form of worry that the neighbors were watching her through her windows. She believed that cars passing in front of her house were driven by spies who were investigating her. She preferred to eat food that came in cans. She was not in any way maternal, but she was an incredible miser. Despite being solidly working class over her lifetime, she had accumulated a small fortune by hoarding whatever money she got from her husband's pensions and her own Social Security checks.

Each of the grandchildren received a portion of her savings, so Mollie's death meant that Arthur and I suddenly had a chunk of money to play with. As Mollie would have liked, we used some of it to pay off bills and put more in savings. We also, going a bit against the grain of how she had lived her own life, repainted the trim on our house and replaced the drafty basement windows. But there was still more money, and around that time we got a flyer in the mail for a liquidation sale on Yamaha grand pianos. My husband is a serious amateur pianist, and for years we have had the upright piano he played throughout his childhood in Omaha. I showed him the flyer about the grand pianos, and he shrugged. I picked up the flyer, called the number listed, and made an appointment for the two of us to try the pianos. "We're doing this together," I said. "I'm going with you, and you're going to look at these pianos." And then he smiled at me, shyly, but happily and said, "That's so nice that you would do that for me."

We never made it to the Yamaha liquidation sale because he found a fabulous piano on eBay instead, a concert grand built in 1899 by Knabe & Co., one of the great piano makers in the United States. He drove three hours to the small Ohio town where the piano was idling in the back of a radiator shop, played it for an hour, then drove the three hours home. That evening he made an offer on the piano, and now it is ours.

Mollie, with all of her frugality, would not have approved of this piano purchase, and I myself might not have just a few months earlier. But now I have Tom in my mind, how one day he was lucid and himself and the next drooling and disoriented. I have Anne in my mind as well, full of life on admission and now completely and irrevocably dead. "Things," life, can change in an instant—I know that now because I have seen it with my own eyes.

Roasting a chicken can soothe many of the bumps in the road that, however tragic they seem at the time, make up ordinary life. But a roasted chicken, or a fricasseed chicken, or a grilled chicken, or even a chicken stir-fried with diamonds, cannot remedy a hurt as deep as death. For that wound, another three words apply: buy the piano.

Arthur plays often now, much more than he used to, and the rich sound of the 110-year-old piano fills our house. The piano gives him a joy like nothing else in his life, and without the piano he would not have known that joy. People say, why wait? But really they should say, don't wait. Listen when you can, tell the people in your life you love them, and buy the piano.

About the Author

Theresa Brown, R.N., lives and works in the Pittsburgh area. She received her B.S.N. from the University of Pittsburgh, and during what she calls her past life, a Ph.D. in English from the University of Chicago. Brown is a regular contributor to the *New York Times* blog "Well." Her essay "Perhaps Death Is Proud; More Reason to Savor Life" was included in *The Best American Science Writing 2009* and *The Best American Medical Writing 2009*. *Critical Care* is her first book. She lives with her husband, Arthur Kosowsky, their three children, and their dog.